Nanotechnology and Occupational Health

Edited by:

Andrew D. Maynard
Woodrow Wilson International Center for Scholars, Washington DC, USA

David Y.H. Pui
University of Minnesota, Minneapolis, MN, USA

Reprinted from the *Journal of Nanoparticle Research*
Volume 9, No. 1, 2007

A C.I.P. catalogue record for this book is available from the Library of Congress

ISBN-13 978-1-4020-5858-5 (HB)
ISBN-13 978-1-4020-5859-2 (e-book)

Published by:
Springer
P.O. Box 17, 3300 AA Dordrecht, The Netherlands

www.springeronline.com

Printed on acid-free paper

Cover illustration: false-color TEM (transmission electron microscopy) image of single walled carbon nanotubes

Journal of Nanoparticle Research

Contents Volume 9 No. 1 February 2007

Special Issue: Nanoparticles and Occupational Health
Guest Editors: Andrew D. Maynard and David Y.H. Pui

Available online
www.springerlink.com

Acknowledgements

Most of the articles in this volume are invited papers from the 2nd International Symposium on Nanotechnology and Occupational Health, held in Minneapolis, MN, October 3-6, 2005, Editors Andrew D. Maynard and David Y.H. Pui, along with other Symposium organizers, would like to specially acknowledge our **major sponsors**:

National Institute for Occupational Safety and Health (NIOSH)
University of Minnesota Office of the Vice President for Research (OVPR)
Center for Biological and Environmental Nanotechnology, Rice University (CBEN)
U.S. Air Force Research Laboratory (AFRL)

Other **co-sponsors** include:

Health & Safety Executive (HSE)
National Nanotechnology Initiative (NNI)
Institute of Occupational Medicine (IOM)
American Society of Mechanical Engineers (ASME)
National Nanotechnology Infrastructure Network (NNIN)
American Industrial Hygiene Association (AIHA)
National Institute of Environmental Health Sciences (NIEHS)
ENVIRON
The Proctor & Gamble Company
National Science Foundation Integrative Graduate Education and Research
Traineeship in Nanoparticle Science and Engineering at the University of Minnesota

The Symposium is facilitated by the College of Continuing Education, University of Minnesota, under the direction of Catherine Ploetz.

Further, Seong Chan Kim, Matthew S. Harrington and David Y.H. Pui, authors of the paper entitled "Experimental study of nanoparticles penetration through commercial filter media" gratefully acknowledge the specific financial support of the National Institute for Occupational Safety and Health (NIOSH).

Journal of Nanoparticle Research (2007) 9:1–3
DOI 10.1007/s11051-006-9164-8

© Springer 2006

Editorial

Special focus: Nanoparticles and Occupational Health

Nanotechnology and occupational health: New technologies – new challenges

Andrew D. Maynard[1,*] and David Y. H. Pui[2]
[1]*Project on Emerging Nanotechnologies, Woodrow Wilson International Center for Scholars, 300 Pennsylvania Avenue NW, Washington, DC, 20004, USA;* [2]*University of Minnesota, Minneapolis, MN, USA;*
*Author for correspondence (Tel.: +1-202-691-4311; Fax: +1-202-691-4001; E-mail: andrew.maynard@wilsoncenter.org)

Received 14 August 2006; accepted in revised form 22 August 2006

Key words: environmental, health and safety, nanotechnology, societal dimensions, occupational health, filtration, exposure measurement, risk analysis

Abstract

An overview of the special issue of the Journal of Nanoparticle Research on nanotechnology and occupational health is presented.

In May 2006, the nanotechnology consultancy firm Lux Research published a report entitled "Taking action on nanotech environmental, safety and health risks" (Lux Research, 2006). The report addressed the potential impact of real and perceived risks on nano-businesses from a commercial perspective, and concluded that "One of the biggest challenges facing firms commercializing nanotechnology innovations today is managing environmental, health and safety (EHS) risks)". At the root of this challenge is uncertainty – including lack of information on real risks, poorly-determined perceptual risks and hesitancy over nanotechnology oversight.

Ultimately, this uncertainty will only be removed through systematic research into what potentially makes engineered nanomaterials hazardous, how this translates into risk, and how both hazard and risk can be managed effectively. No-where is the need for new information more relevant than in nanotechnology laboratories and production facilities; where new materials and products are being designed, produced and used. In 2004, the first international symposium on nanotechnology and occupational health was held in Buxton England, bringing researchers, developers and users together to address critical questions of nanotech health and safety impact (Mark, 2005). Presentations and discussions examined what potentially makes engineered nanomaterials harmful; using existing knowledge to assess and manage risk; the generation of new information for effective risk management; and new ways of working safely with nanomaterials in the face of uncertainty. But perhaps most importantly, the symposium began to break down barriers between different communities, and to encourage research and action that combines expertise from very disparate fields.

The 2004 meeting was highly successful in getting hygienists and manufacturers, toxicologists and materials scientists, and regulators and researchers, talking together. It pointed the way towards what needs to be done, but it also highlighted how little is still known about the potential health risks of some engineered nanomaterials. Twelve months later, and the second international symposium, held in Minneapolis Minnesota, demonstrated how much progress can be made in a short time where there is a will.[1] With nearly three times the attendance of the first meeting and contributions from academics, industry, policymakers, non-government organizations and even lawyers, the second symposium established that, while there is international concern over how to ensure safe nanotech-workplaces, there is also progress being made in developing the knowledge necessary to do this.

This special edition of the Journal of Nanoparticle Research predominantly draws from work presented at the 2005 symposium. It includes new research that will be critical to ensuring worker safety in nanotechnology-industries, as well as highlighting questions that still remain unanswered. The papers are diverse, and cover both original research and new perspectives on nanotechnology and its potential impact within society. The span of topics might seem eclectic at a first glance, and includes particle generation, biological interactions, exposure control and oversight. Yet there is a common thread running through the papers that unites them: The need for a holistic view of nanotechnology and risk that is illuminated by interdisciplinary collaboration, inclusion of diverse perspectives and a need for solutions that are both relevant and workable.

Two themes are dominant in the new research presented here – measurement of airborne nanostructured particles and exposure control. Both are among the top research needs for working safely with engineered nanomaterials. Papers by Fissan et al. and Shin et al. examine the operation and performance of a new instrument, capable of measuring the surface-area of nanometer-scale particles depositing in the gas-exchange region of the lungs. The device, commercially available as the Nanoparticle Surface-Area Monitor (Model 3550, TSI Inc., MN, USA), uses a unique combination of unipolar charging and a charged particle trap, to tailor its response in a way that appears to mimic surface-area dose to the lungs. The surface-area of some inhaled nanoparticles has been shown to be far more relevant than mass concentration in a number of studies (Maynard and Kuempel, 2005), and instruments like this could revolutionize how exposure to such materials is assessed.

There are also five papers examining ways of controlling exposure to airborne nanoparticles. Three of these deserve a special mention, as they address a question that has dogged the handling of engineered nanomaterials in recent times: How effective are filters at removing nanometer–diameter particles from the air? Filter theory, extrapolation of experimental results and even previously published data have so far failed to persuade may people that filters generally are very efficient at collecting nanometer-diameter particles. The papers by Japuntich et al., Wang et al., and Kim et al., represent systematic experimental and theoretical studies of filter efficiency, and should go a long way to allaying fears that filtration is a mechanism of removal not suited to nanoparticles. Importantly, they also show no evidence of the fabled "thermal bounce" –speculated to lead to increased filter penetration at very small particle diameters.

Of course, there are many more questions surrounding the use of filters for airborne nanoparticles than are addressed here – for instance, are low pressure-drop electrostatic filters effective at collecting nanoparticles; do high pressure-drop and bypass leakage become significant factors when using high efficiency filters for nanoparticles; does the most-penetrating particle size shift to the sub-100 nm particle size range for some filter media under certain conditions? Here, as elsewhere, uncertainty persists over how to work as safely as possible with engineered nanomaterials. Yet progress is being made, as demonstrated eloquently by the presentations at the two previous international symposia on nanotechnology and occupational health and the papers presented here. The next international symposium will be held in Taipei Taiwan, in August 2007[2]—the decision to skip 2006 was made intentionally, to reduce the

---

[1] International Symposium on Nanotechnology and Occupational Health: http://www.cce.umn.edu/conferences/nanotechnology/. Accessed August 2005

[2] 3rd International Symposium on Nanotechnology, Occupational and Environmental Health: http://nano-taiwan.sinica.edu.tw/EHS2007/home.htm. Accessed August 2006.

burden of nano-meetings that seems to grow by the month, and to give researchers time to generate new data worth sharing. Our hope is that governments and industry around the world continue to ensure relevant risk-research is appropriately directed and well-funded, and that new research presented at the 2007 symposium will mark a significant reduction in uncertainty over how to assess and manage the risk to health of engineered nanomaterials in the workplace.

References

Lux Research., 2006. Taking action on nanotech environmental, health and safety risks. New York, NY: Lux Research Inc.

Mark D., 2005 Nanomaterials. A risk to health at work?. UK: Buxton UK Health and Safety Laboratory.

Maynard A.D. & E.D. Kuempel, 2005. Airborne nanostructured particles and occupational health. J. Nanopart. Res. 7(6), 587–614.

Journal of Nanoparticle Research (2007) 9:5–10
DOI 10.1007/s11051-006-9193-3

Perspectives
Special issue: Nanoparticles and Occupational Health

Nanotechnology and society

Kenneth H. Keller
*Center for Science, Technology and Public Policy, University of Minnesota, Minneapolis, MN, USA; Author
for correspondence (E-mail: kkeller@jhubc.it)*

Received 3 October 2006; accepted in revised form 16 October 2006

Key words: nanotechnology, society, technology and society interactions, nuclear power, biomedical
engineering, biotechnology, FDA regulatory process

Abstract

Past experience has shown that the successful introduction of a new technology requires careful attention to
the interactions between the technology and society. These interactions are bi-directional: on the one hand,
technology changes and challenges social patterns and, on the other hand, the governance structures and
values of the society affect progress in developing the technology. Nanotechnology is likely to be particularly affected by these kinds of interactions because of its great promise and the unusually early public
attention it has received. Moreover, it represents a new kind of experiment in packaging a rather wide range
of fundamental research activities under a single "mission-like" umbrella. Although this gives it more
impetus as a field, it sets a higher bar for showing successful applications early on and because it links
disparate fields, regulatory regimes reasonable for one kind of nanotechnology development may be
inappropriately extended to others. There are a number of lessons to be gleaned from experience with the
introduction of other technologies, which offer guidance with respect to what pitfalls to avoid and what
issues to be sensitive to as we move forward with the development of nanotechnology applications. The
problems encountered by nuclear power point out the dangers of over-promising and the role the need for
the technology plays in ameliorating fears of risk. The public reaction to biomedical engineering and
biotechnology highlights, in addition, the cultural factors that come into play when technologies raise
questions about what is "natural" and what is "foreign" and what conceptions are involved in defining
"personhood". In all cases, it has been clear that a main task for those introducing new technology is
building public trust–in the safety of the technologies and the integrity of those introducing it. The
advocates of nanotechnology have already shown that they are generally aware of the need to consider the
public's reaction, and they have taken the first steps to act on that awareness. We have to build on those
beginnings, not limiting our considerations simply to issues of safety. If we do so well, we have the
opportunity to develop a new paradigm for technology introduction, which will serve society well in the
future.

Nanotechnology development as a systems problem

In recent years, engineers and scientists have
realized that the dynamics of a complex system
cannot be understood by considering each of its
parts in isolation. In consequence, systems
approaches have become dominant across the
spectrum of the natural sciences and engineering.
In important respects, it is similarly necessary to
take a systems approach to understand the

6

complex interactions between a technology and the society to which it is to be introduced (Guston, 2000).

These interactions are bi-directional. On the one hand, new technologies influence a society's economic and political structures and often raise issues related to the society's values and culture, for example, its concepts of nature, its views of privacy, its attitudes toward personal empowerment and control, and its sense of distributional justice. These factors will often affect how the public perceives the balance of risk and benefit associated with a new technology, particularly where health or the environment is concerned.

On the other hand, the way society structures its policies and institutions for supporting, regulating, and judging the safety of technologies has a strong influence on the pace and direction of their development. Here, too, the relationship is a rather complicated one. First, rapidly changing technologies often render obsolete the existing governmental structures that are intended to support and regulate them: for example, the technology may link previously disparate fields and thus blur the neat separations built into governance regimes; it may introduce new issues that can range from novel intellectual property considerations to the need for new kinds of standards and specifications not previously covered in law; or it may change the relationship between basic research and technology development so that the traditional separation of public and private support responsibilities no longer works. Second, the attitudes that arise from early public judgments of how a technology will affect society (either tangibly or symbolically) will influence the political forces that ultimately determine the policies a government adopts. Thus, early missteps can seriously frustrate a technology's development and prevent society from realizing the benefits it has to offer.

In the past few years, nanotechnology has captured the public attention. Clearly, that is due in large part to its wide range of potential applications, some perhaps more certain than others. However, I think it is fair to say that its public visibility has also resulted from a conscious effort to stir people's imaginations about the promise of the new technology in order to build support for funding the wide range of basic research that might reasonably be included under its rubric, research that is certainly necessary if the applica-

tions are to be successfully developed. Regardless of how one weighs these factors, the public, both those who generally understand what the field is about and those who don't, are clearly engaged and intrigued. But such visibility comes at a price: constant scrutiny; heightened expectations; a reduced willingness to delegate to experts decisions about the use or the safety of the technology; and perhaps a certain skepticism born of the long-standing tension in our society between technophiles and technophobes.

For these reasons it is especially important, as we move forward with this field, to put significant effort into thinking about how the various applications of nanotechnology are likely to play out in practice. The safety of the applications is clearly one of the foci, but for the reasons discussed above, it would be a mistake to limit consideration to that one aspect of the interaction, or to think that such questions can be dealt with entirely by objective risk assessments. We need to think about social, economic, and political factors that affect how a society perceives risk or judges applications; about whether applications are driven by "technology push" or "societal pull"; about whose benefit is balanced against whose risk; about the economic restructuring that can accompany a technological development; about where trust fits in the picture and how it is built (Future R and D Environments, 2002).

We have a good deal of history to draw on in approaching this task. Over the past several decades we have witnessed the problems associated with the introduction of nuclear power for electricity generation (Commission on Engineering and Technical Systems, 1992), the emergence of an extraordinary range of information technology applications with often unanticipated public policy challenges (Computer Science and Telecommunications Board, 2001; Lessig, 1999), and the explosion in applications arising from advances in the biological sciences, ranging from medical devices and tissue engineering (Bronzino, 2006) to genetically-modified organisms (Buttel and Goodman, 2001). What are some of the lessons?

Nuclear power

Although it is still too early to tell how the story will end, particularly given the serious energy problems we confront today, the interesting history of nuclear power illustrates how risk perception can dominate and displace objective risk assessment. A recent study (Sakamoto, 2004) of nuclear power plant performance in the United States, France, and Japan showed that actual plant operating experience with respect to safety and reliability has been very similar in the three countries, yet the public acceptance of the technology has been far different. In France, where close to 80% of the electricity is generated with nuclear plants, and in Japan, where more than 30% comes from nuclear power, there is relatively wide public acceptance and investments in new plants continues. In the U.S., there has not been a new plant started in over 30 years.

It is difficult to account for these differences in any easy way. After all, the arguments against nuclear power, particularly in the U.S., have shifted over time. In the wake of Chernobyl and Three Mile Island, the great concern was a core meltdown and release of radioactivity. Over time, the objections moved to a primary focus on the waste disposal problem. In recent years, that focus has given way to concerns over nuclear terrorism.

Clearly, technical developments have had something to do with the shift in attention, but it is more likely that underlying cause was public skepticism about nuclear power that found expression in these various concerns. And why the skepticism? Three factors seem worth considering. First, the proponents over-promised what the technology could or would achieve. Nuclear-generated electricity would be " ...too cheap to meter" was the oft-quoted statement of a high government official. But, of course, it was not. As with any technology, there were growing problems: cracked cladding in early civilian reactors; huge capital costs in others, unimpressive base plant performance in the early years.

Then came the accidents at Chernobyl and Three Mile Island. And here, again, there was a disconnect between an objective assessment and the public perception. To this day, those two events are linked and, yet, they couldn't be more different. The poor containment design at Chernobyl led to serious escape of radioactivity, which affected a large area of the Ukraine. Containment at Three Mile Island worked exactly as it was intended and the partial core meltdown led to essentially no escape of radioactivity; indeed, it proved the safety of the design (NRC, 2006 a, b). To the public, however, the technical differences between the two situations was not apparent, and they were treated as much the same. Nuclear power is nuclear power, and if it's a problem in one circumstance, it's a problem in all–a kind of "guilt-by-association" extrapolation that may come to haunt what we know to be widely different nanotechnology applications in the future.

Finally, there is the ameliorating factor of need. Neither Japan nor France have significant natural energy resources and their economies would, in the absence of nuclear power, depend heavily on energy imports. The obvious benefit of nuclear power has not only *balanced* risk against need, but also probably *shifted* public perception of risk, as, for example, throughout the world it has shifted our perception of the risk of automobile driving. The very large U.S. coal reserves made nuclear power less necessary and thus removed a factor that might have ameliorated the perception of risk. Interestingly, the increasing concern about climate change appears to be changing that circumstance in the U.S. (Beckjord, 2003).

Biomedical engineering and biotechnology

The effect of the absence of clear need or, perhaps more accurately put, broad benefit, on risk perception has, in many analyst's minds, been a major factor in the early troubles that have been encountered in introducing genetically-modified organisms in agriculture and food. Since the primary beneficiaries of the early agricultural products have been the producers rather than the consumers[1], there was little to balance any concern about risk. Moreover, the social and cultural factors that come into strong play when a society's food supply is involved, evoked a strong emotional response that had a very negative effect on risk perception, leading to headlines about the new "Frankenfoods."

It is interesting that genetic modification, when used in the development of pharmaceuticals or

---

[1]There are strong, but more subtle arguments, that there are broad environmental benefits to genetic plant modifications that reduce the need for insecticides, but they have been muddied by arguments among experts and advocates about more complex ecological considerations.

biologics, has been much more readily accepted by the public than in the case of agricultural or food applications, especially in the U.S. One argument is that the benefit is more apparent and widespread, although in Europe, and particularly in Germany, recombinant DNA technologies even for pharmaceuticals have had a very cool reception because they evoke recollections of eugenics. Of course, strong arguments can be made that recombinant DNA technologies are aimed at helping many to survive who might not otherwise, but what is at work in situations like this is the technology as symbol.

In the U.S., the symbolic aspects of biotechnologies have been a key element in the debates concerning fetal tissue research and applications, therapeutic cloning, and stem cell research. This may explain why scientists who seek to debate these issues solely on technical grounds—such as asserting that the harvesting of fetal tissue can be effectively separated from the decision to abort a fetus; or that the stem cells derived from unused fertility clinic embryos do not destroy an embryo that would otherwise be brought to term; or that using cloning to generate specific differentiated tissues is entirely different from efforts to clone a human being—often find their arguments ignored or failing in the public debate. For the public, a major consideration is the symbolic content of the research or application, the effect the technology has on people's sense of themselves, of their belief systems, of their aesthetic sensibilities. Thus, the former chairman of the President's Council on Bioethics, has put forth the notion of the "yuck factor" or the "wisdom of repugnance" (Kass, 1997) in making judgments about the appropriateness of certain technological developments (Pellegrino, 2006).

This same perspective crops up in the public reaction to medical devices, perhaps even more directly relevant to nanotechnology innovations. Although there are no quantitative studies to corroborate the observation, a number of people have commented that the public is generally more skeptical of and slower to accept therapeutic interventions that involve devices than they are to equally invasive interventions more directly controlled or manipulated by physicians—prescription drugs or surgery, for example. The idea of "mechanical" intervention—a heart valve replacement or an implanted defibrillator or an artificial pancreas, is alienating in its "foreignness", the more so to the general public not in need of such intervention. As a result, failure of such devices is all the more traumatic and unacceptable. For instance, the public would not accept a failure rate in a device that is even close to the failure rate of, say, a chemotherapy regimen. In the early reactions to certain proposed applications of nanoparticles in drug delivery, one can detect a similar skepticism.

One of the ways of gaining the public's confidence is, of course, a rigorous regulatory system that assures the safety and efficacy of a new technology. But here, too, there are some lessons from the medical device field for nanotechnology regulation. Medical devices, like most technologies, develop in an iterative fashion. The experience from their early use feeds back information that leads to the improvement of the technology, lowering its cost and increasing its effectiveness. However, most government regulatory system are unidirectional–gatekeeper approaches that prevent a technology from being put to wide use until it is more or less totally proven and optimally effective. Such a system worked well in the past for the regulation of drugs, but it is inconsistent with the cyclical dynamics of device development. The medical application of nanoparticles is likely to face a similar problem. Therefore, there is a need to develop a staged approach with the early emphasis on safety, subsequent release of the technology for clinical use, and a post-approval emphasis on gathering and feeding back clinical data to scientists and engineers to improve an application's efficacy over time. Thus far, that effort has only been partially successful.

The regulation of medical devices holds one more lesson for those involved in medical applications of nanotechnology. The organization of the FDA regulatory process is based on the assumption that therapeutic systems can be neatly categorized as drugs, biologics, or devices. A separate division exists for each with its own span of control and its own rules and procedures. But, as is often the case, new technologies render old organizational structures obsolete. Combination devices, so called precisely because they combine elements of drugs, biologics and devices, are becoming more common and the existing regulatory framework is problematical (Robinson, 2005). It is easy to envision that drug delivery systems based on nanotechnology will exacerbate the problem by not only crossing the traditional divisional lines, but introducing different classes of regulatory issues within each division, including the need for new ways to categorize and establish performance standards for materials, new migration patterns of materials, and new metabolic interactions involving nanomaterials.

Similarly, we can expect to see a blurring of the boundaries that have traditionally defined the roles of agricultural, environmental, and food and health regulatory agencies. These problems have already arisen in the field of biotechnology with, for example, Bt corn, or the failure to fully separate GMO seed regulation from food product regulation. There are early indications that nanoparticles may present similar issues that complicate the determination of agency responsibility.

Nanotechnology

In examining the lessons of nuclear power and biotechnology, the aim has been to anticipate the kinds of issues that nanotechnology will face as it matures. To scientists it is frustrating to find that the success of a technology does not rest entirely on its objective value, or perhaps more precisely, that its value to a society depends on factors that go well beyond the instrumental or functional. Confidence and trust, symbolism, a balance of risk and benefit (and which groups are affected by each of them), personal autonomy and control are all-important. John Lindsay, a former mayor of New York, once remarked that "New Yorkers are suspicious of breathing air they cannot see." An interesting defense of smog, but putting the humor aside, there is a message, that people need a sense of control of their lives. They need enough knowledge to feel that they are making autonomous choices, and frequently those choices will be based on more than material optimization. We, scientists and engineers, cannot make those choices for them, nor should we believe that our perspective should be entirely dispositive.

In an important respect nanotechnology differs from the example technologies discussed above, and introduces a new and fascinating set of questions. It is arguable that nanotechnology represents an experiment in a new way of generating public support for fundamental research. The examples presented above, as well as many others not discussed here, such as civil aviation development, the space station or the manned voyage to Mars, the "war" on cancer or AIDS, and the early days of the Internet, reflect the political reality that Americans have often had their imaginations fired by focused, or mission-oriented research and development, and they have been willing to support such efforts generously. Basic research, what Vannevar Bush defined as "...the free play of free intellects, working on subjects of their own choice, in the manner dictated by their curiosity for exploration of the unknown," has been the underlying key to mission success, but it has not, itself, fired the public imagination nearly as much. In recent years, particularly in the physical sciences, this has meant flat and, in some years, decreased support for basic research.

Nanotechnology is, in some respects, a packaging of a very broad range of basic research activities into a mission-like activity. Clearly, it has a conceptual framework that has some unity, but the wide range of approaches to synthesis, the variety of mechanisms by which nanoscale properties arise, the different physical manifestations of the products of nanotechnological developments, and the many unrelated applications envisioned for those products, do not suggest a unified field. It would be difficult, for example, to construct a coherent curriculum for nanotechnology (and there would be very little reason for doing so, given the wide reach of the field into other disciplines). It is difficult to imagine that position advertisements for a person with research interests in nanotechnology would have in mind similar kinds of activities in, say, departments of materials science, mechanical engineering, electrical engineering and/or genetics and cell biology, to name but a few.

The nanotechnology approach is a very interesting experiment, precisely because it does address the question of how a nation is best stimulated to support basic research. But it may also amplify some of the problems and challenges of earlier technological developments. First there is the problem of "over-promising". Predicting applications when a science is still relatively early in its research stage risks disappointing a public that is impatient in its expectation of short-term results. Then there is the possibility of "guilt-by-association". When disparate technologies bear a similar label, problems in one are assumed to be extrapolatable to others. If quantum dots in the body create problems by crossing the blood-brain barrier, will the public become concerned that nanofilms in batteries implanted in the body will be a similar threat, although the two have little to do with each other? Finally, there is the question of whether the emphasis on the unity of the field and on early applications will result in the premature development of a regulatory structure that cannot effectively deal with what is an essentially disparate set of technologies.

There are no easy answers to any of these questions, but there are some promising positive signs. Unlike many of the earlier technologies discussed above, the nanotechnology community has moved early and aggressively to address these questions. I am impressed by the inclusion of outreach activities in research grants, and by the creative efforts of informal education venues, such as science centers and museums, to engage the public in discussions that are informative about the technologies and engaging on cost-benefit and values issues. The challenge will be to eschew an approach that focuses entirely on technical questions of safety, even though those questions are vital to address. In a sense, just as the very concept of nanotechnology as a quasi-mission is an experiment, the approach to dealing with the social dimensions of the technology's introduction is another real-time experiment; an attempt to understand how best to balance the necessary delegation of responsibility for the detailed aspects of a new technology to those who are expert in its science, with the need to inform and include the public in the larger questions concerning its use. We all have a stake in doing it well, not only to maximize the promise of nanotechnology's many applications, but to learn more about the dynamics of the science/technology/society interaction that will help us well into the future to manage the course of technology development for the benefit of society.

References

Beckjord E.S., 2003. The Future of Nuclear Power: An Interdisciplinary MIT Study. MIT (http://web.mit.edu/nuclearpower).

Bronzino J.D., 2006. Tissue Engineering and Artificial Organs. Boca Raton: CRC/Taylor and Francis.

Buttel F.H. & R.M. Goodman, 2001. Frankenfoods and Golden Rice : Risks, Rewards, and Realities of Genetically Modified Foods. Madison: Wisconsin Academy of Sciences, Arts and Letters.

Commission on Engineering and Technical Systems., 1992. Nuclear Power: Technical and Institutional Options for the Future. Washington DC: National Academy Press.

Computer Science and Telecommunications Board., 2001. Global Networks and Local Values: A Comparative Look at Germany and the United States. Washington, DC: National Academy Press.

Future R. & D. Environments, 2002. A Report for the National Institute of Standards and Technology. Washington DC: National Academy Press.

Guston D.H., 2000. Between Politics and Science: Assuring the Integrity and Productivity of Research. New York: Cambridge University Press.

Kass L.R., 1997. The Wisdom of Repugnance. Public Broadcasting Service, http://www.pbs.org/wgbh/pages/frontline/shows/fertility/readings/cloning.html.

Lessig L., 1999. Code and Other Laws of Cyberspace. New York: Basic Books.

NRC, 2006 (a). Backgrounder on Chernobyl Nuclear Power Plant Accident. Office of Public Affairs, http://www.nrc.gov/reading-rm/doc-collections/fact-sheets/chernobyl-bg.html).

NRC, 2006 (b). Fact Sheet on the Accident at Three Mile Island. Office of Public Affairs, Washington DC, http://www.nrc.gov/reading-rm/doc-collections/fact-sheets/3mile-isle.html.

Pellegrino E.D., 2006. Advising the President on Ethical Issues Related to Advances in Biomedical Science and Technology. The President's Council on Bioethics, http://www.bioethics.gov.

Robinson R., 2005. US FDA Regulation of Combination Products. Journal of Medical Device Regulation 2(4), 10–19.

Sakamoto H., 2004. Nuclear Power Plant Operating Experiences in France, Japan and the United States. MS Thesis, University of Minnesota, USA.

Journal of Nanoparticle Research (2007) 9:11–22
DOI 10.1007/s11051-006-9173-7

Perspectives
Special issue: Nanoparticles and Occupational Health

Protecting workers and the environment: An environmental NGO's perspective on nanotechnology

John M. Balbus, Karen Florini, Richard A. Denison and Scott A. Walsh
Environmental Defense, 1875 Connecticut Avenue, N.W. Suite 600, Washington, DC, 20009, USA
(Tel.: +1-202-387-3500; E-mail: jbalbus@environmentaldefense.org)

Received 1 August 2006; accepted in revised form 24 August 2006

Key words: nanotoxicology, nanomaterials, nanoscience, nanoparticles, ultrafine particles, health effects, safety, environmental health, occupational health, nanotechnology implications, environmental regulations, occupational regulations

Abstract

Nanotechnology, the design and manipulation of materials at the atomic scale, may well revolutionize many of the ways our society manufactures products, produces energy, and treats diseases. New materials based on nanotechnology are already reaching the market in a wide variety of consumer products. Some of the observed properties of nanomaterials call into question the adequacy of current methods for determining hazard and exposure and for controlling resulting risks. Given the limitations of existing regulatory tools and policies, we believe two distinct kinds of initiatives are needed: first, a major increase in the federal investment in nanomaterial risk research; second, rapid development and implementation of voluntary standards of care pending development of adequate regulatory safeguards in the longer term. Several voluntary programs are currently at various stages of evolution, though the eventual outputs of each of these are still far from clear. Ultimately, effective regulatory safeguards are necessary to provide a level playing field for industry while adequately protecting human health and the environment. This paper reviews the existing toxicological literature on nanomaterials, outlines and analyzes the current regulatory framework, and provides our recommendations, as an environmental non-profit organization, for safe nanotechnology development.

Introduction

Nanotechnology's ability to design and manipulate matter on the atomic scale promises tremendous potential benefits for society, but with significant uncertainties regarding potential damage to the environment and human health. For a set of technologies whose market for applications is expected to reach $1 trillion within 10 years (Roco, 2005b), these uncertainties loom large. Will some nanoparticles persist in the environment and accumulate within living organisms? Do the novel physico-chemical and structural properties of nanomaterials cause unanticipated toxicological behavior at the cellular or organismal levels? Can the potentially harmful properties of nanomaterials be efficiently identified during the development process and engineered out of final products or otherwise effectively managed? These are not just questions of relevance for environmental organizations, those concerned with the health and well-being of workers, or other public health professionals. These questions have been posed by major insurers (e.g., Munich Re, 2002; Swiss Re, 2004; Lauterwasser, 2005) and investment firms (e.g., Langsner et al., 2005; Lux Research, 2005; Wood,

2005) as well as by some major representatives of the burgeoning nanotechnology industry (Denison & Murdock, 2005; Krupp & Holliday, 2005) – all of whom will have to make hard business decisions in the face of these uncertainties. Reducing these uncertainties is critical to ensuring the safe and successful development of nanomaterials, but it will require a greatly increased investment in basic environmental and occupational health research and laboratory testing efforts, as well as an expansion of scientific capacity of regulatory agencies.

Numerous products incorporating nanotechnology are currently on the market, with some, such as cosmetics and clothing, involving clear consumer exposures. (Lux Research, 2005; USEPA, 2005). Many of the products containing nanomaterials that are now on the market have not been subjected to a rigorous review by regulatory agencies, either because they are not required to undergo a pre-market review (e.g., personal care and most consumer products), or because their manufacturer considered them to be essentially the same as existing, bulk substances that are already authorized for use. Other nanomaterials are currently under review by the EPA and FDA. These agencies are by necessity applying regulatory procedures designed for conventional chemical substances and pharmaceuticals to the review of nanomaterials. With commercial development of nanotechnology outpacing the development of a rigorous, comprehensive scientific understanding of the behavior of nanomaterials in biological systems and the potential for human exposures, there is a need to fill gaps in the scientific understanding of potential risks and to develop and implement interim voluntary measures to identify and mitigate those risks.

In the past, commercialization of novel technologies without a thorough assessment of potential risks has led to significant harm to the environment and human health. The widespread use of tetra-ethyl lead in motor fuels impaired the cognitive function of several generations of children, while also possibly decreasing longevity (Lustberg & Silbergeld, 2002) and creating persistent environmental contamination. The lack of attention paid to the harm caused by asbestos resulted in a tremendous human burden of lung disease and mesothelioma. In addition to the human health burden, high litigation and cleanup costs were paid by many of the companies that mined, manufactured and applied asbestos or asbestos-containing products. The total cost of liability for asbestos-related losses is projected to reach $200 billion (Seifert, 2004).

In addition to potential liability concerns, failure to address potential harms and societal concerns proactively, even before widespread health and ecological damage has been demonstrated, could lead to consumer and governmental resistance to new technologies with resulting loss of market share and revenues. This is what befell the emerging biotechnology industry, when European governments' and consumers' resistance to genetically modified foodstuffs is said to have cost the U.S. agricultural sector $200 million in lost crop export revenues in 1998 (Kelch et al., 1998). With nanotechnology development occurring simultaneously in numerous, competing countries around the world, striking the right balance between the development of new, effective, beneficial applications and thorough analysis and management of their potential risks becomes more complex.

Reasons for concerns about risks

Analogies to combustion-related fine and ultrafine particles and early studies of engineered nanoparticles provide some basis for concerns about environmental and health risks from products of nanotechnology. Ultrafine particles, which are in the same size range as nanoparticles, have been demonstrated to traverse the lungs and enter the systemic circulation (Kreyling et al., 2006), where they damage blood vessel linings, hastening atherosclerosis (Kunzli et al., 2005) and leading to a number of other effects. At the cellular level, the small size of ultrafine particles has been shown to allow them to enter and damage cells' mitochondria (Li et al., 2003). There are more numerous studies, using both toxicological and epidemiological techniques, that associate fine (less than 2.5 microns in effective aerodynamic diameter) particles at existing levels of exposure with a wide variety of adverse health outcomes, including heart attacks (D'Ippoliti et al., 2003), strokes (Wellenius et al., 2005), altered electrical activity of the heart (Dockery et al., 2005), lung cancer (Pope et al., 2002), more severe asthma attacks

(Slaughter et al., 2003), stunted lung growth (Gauderman et al., 2004), and possibly decreased intrauterine growth (Wilhelm & Ritz, 2005). Studies of engineered nanoparticles show they also have the ability to traverse lung and even blood-brain (Oberdorster, 2004) and blood-testis (e.g., Chen et al., 2003) barriers. Whether specific types of engineered particles can cause the array of adverse health effects seen with combustion-related fine and ultrafine particles remains to be determined.

In considering analogies such as combustion particles, it is essential to consider the heterogeneity of nanomaterials relative to combustion-generated or other incidentally produced or naturally occurring nanoparticles. Ultrafine combustion particles are heterogeneous both in terms of chemical composition, especially from location to location, and in terms of size and shape. Such particles are generally complex mixtures of combustion-related chemicals adhering to a carbon core. Thus, a single exposure to combustion-related ultrafine particles involves confronting the organism and its component cells with a complex array of particles and chemical contaminants of varying solubility, sizes, and shapes. Any given engineered nanoparticle, on the other hand, while part of a very heterogeneous class of substances, is likely to be far more uniform in terms of size, shape, and chemical composition on an exposure-by-exposure basis. The chemical composition of each type of nanoparticle can range significantly, from metal oxides to linked amines to nearly pure carbon; but particularly from a workplace perspective, the molecular control that is central to nanotechnology typically results in a far narrower shape and size distribution for many nanoparticles. Since shape and size play a large role in determining access to different compartments within the body or even within individual cells, this may mean that a given mass of nanoparticles could consist of a much higher concentration of particles of a specific size and shape. Greater delivery of nanoparticles to specific compartments or cellular organelles could result in greater toxicity compared to more heterogeneous combustion particles. On the other hand, the control over size and shape may also allow re-engineering of nanoparticles to avoid toxicity but still allow function.

Many of the same properties that make nanomaterials uniquely useful in biomedical or other commercial applications could also create novel mechanisms and targets of toxicity. As mentioned above, the ability of certain nanoparticles to penetrate cell membranes, an ability exploited by new applications to deliver targeted therapies, suggests that nanoparticles will also be able to cross physiologic barriers and enter body compartments that larger particles and smaller molecules do not readily access. As another example, carbon-based nanoparticles tend to be extremely strong and durable. Should this durability translate into biopersistence, substances like nanotubes and fullerenes may accumulate in storage sites in the body. Initial studies suggest that carbon nanotubes distribute significantly to the bones (Wang et al., 2004). Short- and long-term effects of this bone accumulation have not been determined. Lastly, several types of nanoparticles, including carbon nanotubes, fullerenes, and dendrimers, are being designed to transport therapeutic agents into specific cells and body compartments (Florence & Hussain, 2001; Kam et al., 2005). In some instances, unanticipated distribution into protected spaces like the nucleus has been noted (Pantarotto et al., 2004). Given the ability of nanotubes (e.g., Zheng et al., 2003; Kam et al., 2005) and fullerenes (in simulations) to bind to and potentially damage DNA (Zhao et al., 2005), this ability to pass through the nuclear membrane is of great concern. And while the focus of researchers has been on these nanoparticles as transporters of therapeutic molecules, the possibility of these molecules also serving as unintended transporters of toxic molecules must be carefully investigated as well.

Increased surface-area-to-mass ratio may be a critical feature in understanding the toxicity of nanomaterials. For a given mass of particles, surface area increases with decreasing particle diameter (and increasing number). In a study comparing the toxicity of conventional vs. nano-sized particles of titanium dioxide, the nanoparticles appeared significantly more toxic when the dose was reported on a mass basis, but the distinction essentially disappeared when the dose was reported on a surface area basis (Oberdorster et al., 2005a). The higher surface-area-to-mass ratio also leads to higher particle surface energy, which may translate into higher reactivity (Oberdorster et al., 2005b).

Lastly, the combination of high surface area and small size may give nanoparticles unusual catalytic reactivity due to quantum effects, such as those seen with gold nanoparticles (Daniel & Astruc, 2004). This combination of enhanced surface area and enhanced surface activity lends far greater complexity to the characterization of nanoparticles, and also precludes simple extrapolation of toxicity among nanoparticles of different sizes and surface chemistry.

Surface modifications may allow nanoparticles to bind to cell surface receptors and either avoid internalization (Gupta & Curtis, 2004) or be taken up by specific transport mechanisms, allowing cell targeting for therapeutic agents. It is clear that even subtle variations in nanoparticle surfaces, whether due to intentional coating prior to entry into the body, unintentional surface binding of proteins, or degradation of coatings once inside the body, can have dramatic impacts on where and how nanoparticles gain entry into cells, as well as where and how they are transported within cells after entry. Understanding the implications of surface modifications as well as assuring the stability of surface properties throughout the lifespan of manufactured nanoparticles will be critical to assuring safety.

The database of toxicity studies on nanomaterials is extremely limited. While several long-commercialized substances containing nanoparticles, such as carbon black and titanium dioxide, have undergone chronic bioassays (Nikula et al., 2001), there have been no published studies to date examining chronic health effects of newer, highly engineered nanoparticles (International Council On Nanotechnology, 2006). Virtually all of the studies done to date examine only short-term effects; many are limited to *in vitro* tests of cultured cells. Also, current research funding does not appear to be examining chronic health effects, such as cancer or developmental effects (Woodrow Wilson Center's Project on Emerging Nanotechnologies, 2006). Of the limited number of short-term studies completed to date, several have found a variety of adverse effects. Studies in which single-walled carbon nanotubes (SWCNTs) were instilled or aspirated into the lungs of rodents have consistently demonstrated that SWCNTs cause lung granulomas and other signs of acute lung inflammation (Lam et al., 2003; Warheit et al., 2004; Shvedova et al., 2005) and one (Shvedova et al.,

2005) found that SWCNTs also cause dose-dependent, diffuse interstitial fibrosis. Similar effects were seen in one study of multi-walled carbon nanotubes (MWCNTs) (Muller et al., 2005). The finding of diffuse fibrosis is especially concerning for its potential to impair lung function.

C_{60} fullerenes (commonly known as buckyballs) have been less well-studied in mammalian models. They have been shown to be potent bactericides in water (Fortner et al., 2005). A second study purports to demonstrate transport via the gills from water to the brains of fish, with subsequent oxidative damage to brain cell membranes (Oberdorster, 2004). Uncoated buckyballs also have caused oxidative stress in *in vitro* testing systems, although hydroxylated and other derivatized buckyballs appear to protect against oxidative stress in biological systems (Sayes et al., 2004, 2005). Some authors have questioned whether observed toxicity from fullerenes is instead caused by organic solvents contaminating the aqueous fullerene colloids. They point to other studies (including *in vivo* studies) that show negligible toxicity and even protective effects from pristine fullerenes that are made into water-soluble aggregates without the use of organic solvents (Andrievsky et al., 2005; Gharbi et al., 2005). Further studies are needed to resolve these discrepancies.

Quantum dots can be made of a variety of inherently toxic materials, including cadmium and lead. As some of the key applications of quantum dots include diagnostic imaging and medical therapeutics, quantum dots have been studied relatively extensively in biological systems, although only a small portion of this research has focused directly on potential toxicity. Toxicological studies performed to date have mainly been *in vitro* cytotoxicity assays. While results have been somewhat inconsistent, studies that used longer exposure times were more likely to demonstrate significant toxicity (Hardman, 2005). Quantum dots typically have a core made of inorganic elements, but they are generally coated with organic materials such as polyethylene glycol to enhance their biocompatibility or target them to specific organs or cells. Some coatings initially decrease toxicity by one or more orders of magnitude, but the coatings are known to degrade when exposed to air or ultraviolet light, after which toxicity increases. While the presumption has been that this cytotoxicity was caused by leakage of cadmium or selenium from the core, there is evidence that some of the molecules used as coatings may have independent toxicity (Hardman, 2005). Significant questions remain about the safety of quantum dots based on the available *in vitro* studies.

How well will current regulatory frameworks protect workers, the public and the environment from nanomaterial risks?

Nanotechnology will challenge current occupational and environmental regulatory frameworks for a number of reasons. First, in most current regulatory programs, standards (and exemptions from them) are based on mass and mass concentration. Because of their high surface-area-to-mass ratios and enhanced surface activity, some nanomaterials are likely to prove potent at far lower concentration levels than those envisioned when threshold standards were initially set. Second, regulators often rely on structure–activity models to extrapolate and predict at least some types of toxicity for new conventional materials. Too little is currently known about nanomaterials to enable such extrapolation.

Third, it appears many nanomaterials are being developed by small, start-up businesses, which tend to focus on a small number of products. By April 2006, there were approximately 1500 start-ups focused on nanotechnology worldwide (Lux Research Inc., 2006). As a result, knowing exactly which materials are being produced and used, by what processes and for what applications – and directing any compliance and enforcement efforts to where they are needed – will be hampered by the sheer number of facilities involved. By the same token, a great deal of production, processing and use will take place in facilities that may lack expertise and resources to understand and comply with environmental and occupational safeguards.

Lastly, the pace of the regulatory process lags far behind the speed with which nanomaterials are being brought to market. While substances marketed as drugs, food additives, fuel additives, and pesticides typically receive significant scrutiny when first brought to market, most others do not. As a result, occupational and environmental protections generally are developed only after problems are identified or strongly suspected, and then

in regulatory proceedings that typically take many years to complete. The opportunity exists to recognize and control problems more proactively with nanotechnology.

A more detailed discussion of specific regulatory issues follows.

Occupational Safety and Health Administration

As of June 2006, the Occupational Safety and Health Administration (OSHA) had not published any standards, guidance, or position papers on nanotechnology. While the agency does participate in the National Nanotechnology Initiative (NNI), it is unclear what nanotechnology-specific activities are underway at the time of writing. On the other hand, the non-regulatory National Institute of Occupational Safety and Health (NIOSH) has developed several useful draft guidance documents regarding occupational safety and health practices for the nanotechnology industry (see, e.g., NIOSH, 2005). These documents address health and safety concerns, exposure monitoring, engineering controls, and workplace practices for nanotechnology manufacturing facilities. Presently, they do not constitute official guidance, but are draft documents open for public comment.

Under the Occupational Safety and Health Act (OSHAct), four types of standards are relevant for protecting workers from overexposure to nanomaterials: substance-specific standards, general respiratory protection standards, the hazard communication standard, and the "general duty clause." Each is examined below.

Given the slow pace at which toxicity data on nanoparticles are being developed, as well as the historically slow pace and high hurdles facing occupational standard-setting, it is unlikely that any nanoparticle-specific standards will be put in place in the next several years. In the absence of specific standards, inhalable nanoparticles will only be addressed by the 5 mg/m^3 standard that applies to "particulates not otherwise regulated", sometimes called "nuisance dust" (29 CFR section 1910.1000 Table Z-1). These mass-based standards, developed for conventional particles, may not protect workers from adverse effects of chronic nanoparticle exposures. While extrapolation from instillation studies is problematic, the concentrations used in studies finding lung granulomas and inflammation in rats and mice exposed to carbon nanotubes are equivalent to that which a worker exposed at 5 mg/m^3 would receive within several weeks (Lam et al., 2003; Shvedova et al., 2005). In the absence of rigorous, science based standards that address the unique aspects of nanomaterials, protection of nanotechnology workers will depend upon voluntary precautionary measures on the part of industry, with a weak backstop provided by OSHA's general duty clause (see below). The rapid development of toxicological information and environmental fate and transport knowledge on a representative set of nanomaterials would be very helpful in informing the occupational health and safety staff at the companies who must design and decide upon such voluntary measures.

The respiratory protection standard (29 CFR section 1910.134) requires employers to provide workers with respirators or other protective devices when engineering controls are not adequate to protect health. The standard provides guidance in selecting specific personal protective equipment and in implementing workplace respiratory protection programs. Only respirators certified by the National Institute of Occupational Safety and Health (NIOSH) may be used, and employers must assess the effectiveness of the respirators they supply. The current lack of validated means to measure and characterize the form and size of nanoparticles in the air, as well as uncertainties regarding respirator performance, especially with particles between 30 and 70 nanometers and potential agglomerates around 300 nanometers (Balazy et al., 2006), will complicate implementation of this standard.

Third, OSHA's hazard communication standard (CFR section 1910.1200) stipulates that all producers or importers of chemicals are obligated to develop material safety data sheets ("MSDSs"), which are intended to provide workers with available information on hazardous ingredients in products they handle and educate them on safe handling practices. However, even when accurate and up-to-date, MSDSs have significant limitations – most notably, there is no requirement either to generate data on potential hazards, or to disclose the absence of data. Moreover, in some instances a nanomaterial's MSDS has simply adopted the hazard profile for a presumed-related bulk material. For example, an MSDS for carbon nanotubes identifies the primary component as graphite, and goes on to cite information on the hazards of graphite without acknowledging any dissimilarity between the two substances (Carbon Nanotubes, Inc., 2004).

Finally, OSHAct's general duty clause (section 5(a)(1), 29 USC section 654) is intended as a backstop to protect workers from exposures that are widely known to result in toxic effects but are not addressed specifically by an OSHA standard. The general duty clause, however, applies only to "recognized" hazards, a difficult criterion to meet in light of the current paucity of toxicity data on specific nanomaterials.

Environmental Protection Agency

The Environmental Protection Agency conducts both regulatory and research activities relevant to protecting the general public and the environment from potential risks of nanotechnology. The agency's current thinking has been summarized in a draft Nanotechnology White Paper released in December 2005 (USEPA, 2005). The white paper summarizes hazard- and exposure-related concerns as well as environmental applications of greatest interest to the EPA. It also describes the range of regulatory authorities under EPA that may ultimately be relevant to nanotechnology. These include the Clean Air Act, Clean Water Act, the Federal Insecticides and Rodenticides Act (FIFRA), the Resource Conservation and Recovery Act (RCRA, which addresses management of hazardous and other solid wastes), and the Toxic Substances Control Act (TSCA, which covers chemicals other than drugs, food additives, cosmetics, and pesticides). The white paper notes that the agency has already received notices of the intention to manufacture nanomaterials pursuant to the provisions of TSCA that govern new chemicals, as well as a request for approval of a fuel additive under the Clean Air Act. However, the white paper does not indicate in any detail what information the agency will use, or how it will obtain it, in order to make decisions on these applications. These issues are discussed in greater detail below. The white paper concludes with recommendations for integrating nanotechnology into its pollution prevention programs; an ambitious research program on environmental applications and environmental and health implications

17

of nanotechnology; a cross-agency coordinating workgroup; case studies on risk assessment; and training needs. It contains no recommendations, however, for initiating regulatory action.

New nanomaterials will come under the purview of TSCA. Section 5 of TSCA requires the producer of a "new" chemical substance to send EPA a "Pre-Manufacture Notification" (PMN) 90 days before beginning to produce a substance. Unfortunately, there are no baseline data requirements for PMNs, and 85% of PMNs received by EPA for conventional chemicals are submitted without any health data (Government Accountability Office, 2005). EPA can request additional data, but rarely does so; it typically conducts its review based on use of structure-activity relationship models, through which toxicological properties of an unstudied substance are estimated based on the extent of molecular structural similarity to substances with known toxicological properties. As noted in the white paper, however, the existing models have little applicability to nanomaterials. This is because the models are based on the properties (primarily molecular structure) of bulk forms of conventional chemical substances, whereas nanomaterials' novel and enhanced properties result from characteristics (e.g., size, shape) in addition to their molecular structure. It remains to be seen whether, in the absence of an existing knowledge base and predictive models for nanomaterials, the EPA will routinely require actual toxicity data to be generated and included in PMN submissions.

Other key questions also remain unresolved, including the extent to which nanomaterials qualify as "new" chemicals (necessary to trigger PMN requirements). Under TSCA, a "new" chemical is one that is not already listed on the TSCA Inventory of chemicals in commerce, and a chemical is defined as a substance with "a particular molecular identity" (TSCA section 3, 15 USC section 2602(2)). While nanomaterials whose molecular formula is not already included on the TSCA Inventory obviously constitute "new" materials, some parties appear to be assuming that other nanomaterials – those with a molecular formula identical to a substance already on the Inventory – do not qualify as new.

TSCA also provides certain information-gathering authorities. Under Section 8(a), EPA can require manufacturers to provide certain use and exposure information. Section 8(e) requires manufacturers to submit any information indicating that a substance may pose a "significant risk" to health or the environment, while Section 8(d) allows EPA to require manufacturers to submit all toxicity-related studies already in their possession. EPA has indicated an intention to exercise its regulatory authority to gather information, and is conducting a multi-stakeholder process that is both designing a voluntary initiative to address nanomaterial risks and considering possible use of TSCA authorities (National Pollution Prevention and Toxics Advisory Committee (NPPTAC), 2005).

Under section 211 of the Clean Air Act, new fuel additives must be registered with the EPA, with manufacturers required to supply EPA with certain data to allow for a product safety assessment. Manufacturers are required to measure and speciate their emissions and submit a literature review cataloguing any known health effects of the substances being emitted. There is also a second level of testing requirements mandating toxicological studies on animals. However, small businesses with annual revenues of less than $10 million are exempt from these requirements. This is reason for concern with respect to nanotechnology, since many products, including those used in highly dispersive applications like fuel additives, are being developed by smaller companies that qualify for this exemption. For example, a diesel fuel additive utilizing cerium oxide nanoparticles has been submitted for approval to EPA. The company that submitted the request for approval, Oxonica (through its Cerulean division), currently qualifies for the small business exemption from more rigorous testing, even though it is partnering with the multinational corporation BASF in its commercialization and international marketing of the additive (AzoNano.com, 2004). While EPA has authority to still require testing if it decides there are insufficient data to assure safety, the burden is on the agency to justify such testing.

Addressing nanomaterial risks: Next steps

Safe and responsible development of nanotechnology thus presents a number of challenges. These include ensuring thorough and timely evaluation of nanomaterials prior to commercialization; balancing the benefits as well as the unknown risks of new nanomaterials with the sometimes better-known risks of substances they would be replacing; and applying appropriate safeguards to the production, use, and disposal of engineered nanomaterials in the face of the uncertainties listed above.

Given the limitations of existing regulatory tools and policies, we believe two distinct kinds of initiatives are needed now: first, a major increase in the federal investment in nanomaterial risk research, and second, rapid development and implementation of voluntary standards of care, pending development of adequate regulatory safeguards. A wide array of stakeholders must be involved in all components of the latter process, not only large and small businesses and the academic community, but also labor groups, health organizations, consumer advocates, community groups, and environmental organizations.

Increase governmental investment in risk research

The U.S. government, as the largest single investor in nanotechnology research and development, needs to spend more to assess the health and environmental implications of nanotechnology and ensure that the critical research needed to identify potential risks is done expeditiously. Through the National Nanotechnology Initiative, the federal government spends about $1.3 billion annually on nanotechnology research and development. Initial efforts to fund studies of the environmental health and safety (EHS) and ethical, legal, and social implications (ELSI) issues pertaining to nanotechnology were led by the NSF starting in 2001 (Roco, 2005a). Current funding is relatively limited. The NNI indicates that its spending on research "whose *primary* purpose is to understand risk" amounts to $44.1 million for FY07, or less than 3.5% of total NNI funding (NSET, 2006). Even this figure may be optimistic: the Woodrow Wilson International Center for Scholars' Project of Emerging Nanotechnologies (PEN) has found that funding of "highly relevant" nanotechnology risk research was less than 1% of the 2005 annual NNI budget, totaling about $11 million (Maynard, 2006). The funding for the broader category of research PEN deemed "relevant" to health and safety risks of nanotechnology was estimated to be $31 million, less than 3% of

the 2005 NNI budget (Maynard, 2006). Both estimates are considerably smaller than the nearly $40 million claimed by NNI to have been spent on EHS research that year.

We recommend that the U.S. government should spend at least $100 million annually on hazard and exposure research for at least the next several years. Given the complexity of the task, the scope of the necessary research, and available benchmarks for comparison, $100 million per year represents a reasonable lower-bound estimate of what is needed (Denison, 2005). The need for a substantial increase in risk research is supported by numerous expert assessments. For example, invited experts to a workshop sponsored by the Nanoscale Science Engineering, Science and Technology Subcommittee (NSET) of the NNI called for at least a 10-fold increase in federal spending on nanotechnology risk-related research, relative to the approximately $10 million spent in FY2004 (Phibbs, 2004). Additionally, President Bush's science advisor John H. Marburger III noted that the current toxicity studies now under way through the NNI are "a drop in the bucket compared to what needs to be done" (Weiss, 2005). One can also look to other test batteries to gauge the approximate cost for health and environmental testing for nanotechnologies. The hazard-only test battery required of pesticides under FIFRA provides a good example. The Agricultural Research Service estimates that this test battery, which consists of up to 100 individual data elements (40 CFR Part 158) and might be initially appropriate for testing of some nanomaterials, can cost up to $10 million per chemical for a pesticide proposed for major food crop use (U.S. Department of Agriculture Agricultural Research Service, 1997). An additional benchmark for judging the appropriate level of federal expenditure for nanomaterial risk research is the budget for EPA research on risks posed by airborne particulate matter recommended by the National Research Council in 1998 – the scope of which was considerably narrower than the needed research on nanomaterials. The recommended budgets, and subsequent EPA expenditures, ranged between $40–60 million annually for the first six years (NRC, 1998, 2004).

Taken together, these benchmarks indicate that at least $100 million annually over a number of years is a justifiable amount for the federal government to invest in health and safety research in order to address the major unknowns and uncertainties associated with the burgeoning field of nanotechnology. It should be noted that this figure is quite small in comparison to the $1 trillion role that nanotechnology is projected to play in the world economy by 2015.

But the U.S. government need not be the sole, or even the principal, funder of nanomaterial risk research. Other countries are also spending heavily to promote nanotechnology research and development, and they too should allocate some portion of their spending to address nanotechnology risks. Coordination of such investment, perhaps through organizations such as the Organization for Economic Cooperation and Development (OECD), is essential. And although government risk research has a critical role to play in developing the basic knowledge and methods to characterize and assess the risks of nanomaterials, private industry should fund the majority of the research and testing on the products they are planning to bring to market. Clearly, all parties will benefit if governments and industry coordinate their research to avoid redundancy and optimize efficiency.

Develop voluntary standards of care

Because federal agencies may not put into place adequate provisions for nanomaterials quickly enough to address the products now entering or poised to enter the market, voluntary "standards of care" for nanomaterials must play a role in guiding the safe use of nanomaterials in the meantime. These standards should include a framework and a process by which to identify and manage nanomaterials' risks across a product's full lifecycle, taking into account worker safety, manufacturing releases and wastes, product use, and product disposal. Such standards should be developed and implemented in a transparent and accountable manner, including public disclosure of the assumptions, processes, and results of the risk identification and risk management systems.

Several voluntary programs are currently at various stages of evolution, though the eventual outputs of each of these are still far from clear. As noted above, in October 2005, a workgroup of an EPA advisory committee proposed a framework for a voluntary program aimed at producers, processors, and users of nanomaterials. The group also recommended using certain TSCA regulatory authorities to address nanomaterial risks (National Pollution Prevention and Toxics Advisory Committee (NPPTAC), 2005). In addition, Environmental Defense is working directly with DuPont to develop a framework for the responsible development, production, use, and disposal of nano-scale materials (Environmental Defense, 2005). While the project will initially pilot-test the framework on specific nano-scale materials or applications of interest to DuPont, the organizations intend to develop a framework that can be adapted for use by a broad range of stakeholders. Other multi-stakeholder efforts to develop voluntary standards are also underway through ASTM International (ASTM International, 2005) and the International Standards Organization, which recently convened a new Technical Committee on Nanotechnologies (International Standards Organization, 2005).

In the long run, regulatory programs will be essential to securing long-term public confidence in and support for nanotechnology (Macoubrie, 2005) as well as leveling the playing field among large and small businesses bringing new products incorporating nanotechnology into the market. Because commercialization is taking place before research is able to resolve fundamental uncertainties on the behavior and safety of nanomaterials, there is an urgent need for both toxicity testing for new products as well as workplace and environmental controls to minimize the possibility of exposures. Voluntary programs can be useful for developing an understanding of how such measures can be instituted, but ultimately, to ensure equitable application of principles of safe development, a strong regulatory framework will be required.

Conclusion

Nanotechnology holds the potential to help achieve cleaner air, water, and soil, more effectively treat disease, and improve energy efficiency and material durability. Many of the same physico-chemical properties that give nanomaterials so much promise, however, leave open the possibility that they could have adverse effects on human health and the environment. We believe the combination of existing scientific knowledge, as well as the recent memories of unintended consequences of other technological advances, provide sufficient motivation for nanotechnology industries and relevant government agencies to invest in understanding potential risks and either engineering them out of materials and products before commercialization or effectively managing them from the start in other ways. Public health and public trust will both be maximized by proactive efforts to get nanotechnology right the first time.

References

Andrievsky G., V. Klochkov & L. Derevyanchenko, 2005. Is C_{60} fullerene molecule toxic? Fuller. Nanotub. Car. N 13, 363–376.

ASTM International, 2005. Committee E56 on Nanotechnology [Internet]. [cited 2006 July 5]. Available from: http://www.astm.org/COMMIT/COMMITTEE/E56.htm.

AzoNano.com. Oxonica and BASF to Collaborate on Commercialisation of Fuel Additive [Internet], 2004 July 13 [cited 2006 July 5]. Available from: http://www.azonano.com/news.asp?newsID = 221.

Balazy A., M. Toivola, T. Reponen, A. Podgorski, A. Zimmer & S. Grinshpun, 2006. Manikin-based performance evaluation of N95 filtering-facepiece respirators challenged with nanoparticles. Ann. Occup. Hyg. 50, 259–269.

Carbon Nanotubes, Inc. Material Safety Data Sheet – CNI Carbon Nanotubes [Internet], 2004 September 17 [cited 2006 July 5]. Available from: http://www.cnanotech.com/download_files/MSDS%20CNI%20Nanotubes.pdf.

Chen Y., Z. Xue, D. Zheng, K. Xia, Y. Zhao, T. Liu, Z. Long & J. Xia, 2003. Sodium chloride modified silica nanoparticles as a non-viral vector with a high efficiency of DNA transfer into cells. Curr. Gene Ther. 3(3), 273–279.

Code of Federal Regulations, 29 CFR §1910.134:2005.

Code of Federal Regulations, 29 CFR §1910.1000:2005. Table Z-1.

Code of Federal Regulations, 29 CFR §1910.1200:2005.

Code of Federal Regulations, 40 CFR § 158:2005.

Daniel M. & D. Astruc, 2004. Gold nanoparticles: assembly, supramolecular chemistry, quantum-size-related properties, and applications toward biology, catalysis, and nanotechnology. Chem. Rev. 104, 293–346.

Denison R., 2005. A proposal to increase federal funding of nanotechnology risk research to at least $100 million annually [Internet], 2006 [cited 2006 July 5]. Available from: http://www.environmentaldefense.org/documents/4442_100 milquestionl.pdf.

Denison R. & S. Murdock, 2005 November 17. An ounce of prevention. Fort Wayne News Sentinel and other Knight Ridder/Tribune News Service newspapers.

D'Ippoliti D., F. Forastiere, C. Ancona, N. Agabiti, D. Fusco, P Michelozzi & C. Perucci, 2003. Air pollution and myocardial infarction in Rome: a case-crossover analysis. Epidemiol. 14(5), 528–535.

Dockery D., H. Lettmann-Gibson, D. Rich, M. Link, M. Mittleman, D. Gold, P. Koutrakis, J. Schwartz & R. Verrier, 2005. Association of air pollution with increased incidence of centricular tachyarrhythmias recorded by implanted cardioverter defibrillators. Environ. Health Perspect. 113(6), 670–674.

Environmental Defense, 2005 [cited 2006 July 5]. Environmental Defense and DuPont: Global Nanotechnology Standards of Care Partnership [Internet]. Available from: http://www.environmentaldefense.org/article.cfm?contentID = 4821.

Florence A. & N. Hussain, 2001. Transcytosis of nanoparticle and dendrimer delivery systems: evolving vistas. Adv. Drug Deliv. Rev. Oct 1; 50 Suppl 1, S69–S89 (Review).

Fortner J., D. Lyon, C. Sayes, A. Boyd, J. Falkner, E. Hotze, L. Alemany, Y. Tao, W. Guo, K. Ausman, V. Colvin & J. Hughes, 2005. C60 in water: nanocrystal formation and microbial response. Environ. Sci. Technol. 39, 4307–4316.

Gauderman J., E. Avol, F. Gilliland, H. Vora, D. Thomas, K. Berhane, R. McConnell, N. Kuenzli, F. Lurmann, E. Rappaport, H. Margolis, D. Bates & J. Peters, 2004. The effect of air pollution on lung development from 10 to 18 years of age. NEJM 351(11), 1057–1067.

Gharbi N., M. Pressac, M. Hadchouel, H. Szwarc, S. Wilson & F. Moussa, 2005. Fullerene is a powerful antioxidant in vivo with no acute or subacute toxicity. Nano. Lett. 5(12), 2578–2585.

Government Accountability Office, 12 June 2005. "Options Exist to Improve EPA's Ability to Assess Health Risks and Manage Its Chemical Review Program," GAO-05-458, p. 11.

Gupta A. & A. Curtis, 2004. Lactoferrin and ceruloplasmin derivatized superparamagnetic iron oxide nanoparticles for targeting cell surface receptors. Biomaterials. 25, 3029–3040.

Hardman R., 2005. A toxicological review of quantum dots: toxicity depends on physico-chemical and environmental factors. Environ. Health Persp. Nat. Inst. of Environ. Health Sci. doi: 10.1289/ehp.8284.

International Council on Nanotechnology (ICON), 2006 [cited 2006 July 5]. EHS Database [Internet]. Available from: http://icon.rice.edu/research.cfm.

International Standards Organization, 2005 [cited 2006 July 5]. Nanotechnologies Technical Committee – TC 229 [Internet]. Available from: http://www.iso.org/iso/en/stdsdevelopment/tc/tclist/TechnicalCommitteeDetailPage.TechnicalCommitteeDetail?COMMID = 5932.

Kam N., M. O'Connell, J. Wisdom & H. Dai, 2005. Carbon nanotubes as multifunctional biological transporters and near-infrared agents for selective cancer cell destruction. Proc. Natl. Acad. Sci. 102(33), 11600–11605.

Kelch D.R., M. Simone & M.L. Madell, 1998 [cited 2006 August 2]. U.S. Department of Agriculture Economic Research Service. Biotechnology in Agriculture Confronts Agreements in the WTO [Internet]. Available from: http://www.ers.usda.gov/publications/wrs984/wrs984e.pdf.

Kreyling W.G., M. Semmler-Behnke & W. Moller, 2006. Ultrafine particle-lung interactions: does size matter? J. Aerosol Med. Spring 19(1), 74–83.

Krupp F., & C. Holliday, 2005 June 14; page. Let's Get Nanotech Right. Wall Street Journal, p. B2.

Kunzli N., M. Jerrett, W. Mack, B. Beckerman, L. LaBree F. Gilliland, D. Thomas, J. Peters & H. Hodis, 2005. Ambient air pollution and atherosclerosis in Los Angeles. Environ. Health Perspect. 113(2), 201–206.

Lam C., J. James, R. McCluskey & R. Hunter, 2003. Pulmonary toxicity of single-wall carbon nanotubes in mice 7 and 90 days after intratracheal instillation. Toxicol. Sci. 77, 126–134.

Langsner H., S. Martinez, D. Zaveri, K. Iguchi, R. Sumangali & M. Milcetich, 2005. Nanotechnology: Non-traditional methods for valuation of nanotechnology producers. Report Prepared by Innovest Strategic Value Advisors. New York, NY.

Lauterwasser C. ed. 2005. Allianz Group. Small sizes that matter: Opportunities and risks of Nanotechnologies. Report in co-operation with the OECD International Futures Programme. Munich, Germany.

Li N., C. Sioutas, A. Cho, D. Schmitz, C. Misra, J. Sempf M. Wang, T. Oberley, J. Froines & A. Nel, 2003. Ultrafine particulate pollutants induce oxidative stress and mitochondrial damage. Environ. Health Perspect. 111(4), 455–460.

Lustberg M. & E. Silbergeld, 2002. Blood lead levels and mortality. Arch. Intern. Med. 162, 2443–2449.

Lux Research Inc., 2006. The Nanotech Report, 4th edn. New York (NY).

Lux Research Inc., 2005. A prudent approach to nanotechnology, environmental, health, and safety risks. New York (NY).

Macoubrie J., 2005 Informed Public Perceptions of Nanotechnology and Trust in Government. Washington, DC: Woodrow Wilson International Center for Scholars.

Maynard A.D., 2006 Nanotechnology: A Research Strategy for Addressing Risk. Washington, DC: Woodrow Wilson International Center for Scholars.

Muller J., F. Huaux, N. Moreau, P. Mission, J. Heilier M. Delos, M. Arras, A. Fonseca, J. Nagy & D. Lison, 2005. Respiratory toxicity of multi-wall carbon nanotubes. Toxicol. Appl. Pharmacol. 207, 221–231.

Munich Re Group., 2002. Nanotechnology: What is in Store for Us? Munich, Germany: Munich Re Group.

National Institute of Occupational Safety and Health, 2005. Approaches to Safe Nanotechnology: An Information Exchange with NIOSH [Internet]. [cited 2006 July 5] Available from: http://www.cdc.gov/NIOSH/topics/nanotech/nano_exchange.html.

National Pollution Prevention and Toxics Advisory Committee (NPPTAC), advisory committee to the U.S. Environmental Protection Agency, 22 November 2005. Interim Ad Hoc Work Group on Nanoscale Materials. Overview of Issues for Consideration by NPPTAC. Available from: http://www. regulations.gov (use Advanced Search's Document Search function, enter EPA-HQ-OPPT-2002–0001–0068 as the Document ID).

Nikula K.J., V. Vallyathan, F. Green & F. Hahn, 2001. Influence of exposure concentration or dose on the distribution of particulate material in rat and human lungs. Environ. Health Perspect. 109, 311–318.

NRC (National Research Council)., 1998. Research Priorities for Airborne Particulate Matter: 1. Immediate Priorities and a Long-Range Research Portfolio. National Academy Press: Washington, DC.

NRC (National Research Council)., 2004. Research Priorities for Airborne Particulate Matter: 4. Continuing Research Progress. Washington, DC: National Academy Press.

Nanoscale Science, Engineering and Technology (NSET), Subcommittee of the Committee on Technology, National Science and Technology Council, 2005 March [cited 2006 July 5]. The National Nanotechnology Initiative: Research and Development Leading to a Revolution in Technology and Industry: Supplement to the President's FY2006 Budget [Internet]. Available from: http://www.nano.gov/NNI_06Budget.pdf.

Oberdorster E., 2004. Manufactured nanomaterials (fullerenes, C60) induce oxidative stress in the brain of juvenile largemouth bass. Environ. Health Perspect. 112, 1058–1062.

Oberdorster G., E. Oberdörster & J. Oberdörster, 2005a. Nanotoxicology: an emerging discipline evolving from studies of ultrafine particles. Environ. Health Perspect. 113, 823–839.

Oberdorster G., A. Maynard, K. Donaldson, V. Castranova J. Fitzpatrick, K. Ausman, J. Carter, B. Karn, W. Kreyling, D. Lai, S. Olin, N. Monteiro-Riviere, D. Warheit & H. Yang, 2005b. Principles for characterizing the potential human health effects from exposure to nanomaterials: elements of a screening strategy. Part. Fibre Toxicol. 2, 8 .

Occupational Safety and Health Act of 1971, § 5:29 U.S.C. § 654.

Pantarotto D., J. Briand, M. Prato & A. Bianco, 2004. Translocation of bioactive peptides across cell membranes by carbon nanotubes. Chem. Commun. (Camb). 1, 16–17.

Phibbs P., 13 August 2004. Federal Government Urged to Boost Spending on Managing Risks Posed by Nanotechnology. Daily Environment Report, p. A-3.

Pope C., R. Burnett, M. Thun, E. Calle, D. Krewski, K. Ito & G. Thurston, 2002. Lung cancer, cardiopulmonary mortality, and long-term exposure to fine particulate air pollution. JAMA 287(9), 1132–1141.

Rejman J., V. Oberle, I. Zuhorn & D. Hoekstra, 2004. Size-dependent internalization of particles via the pathways of clathrin- and caveolae-mediated endocytosis. Biochem. J. 377, 159–169.

Roco M.C., 2005a. The emergence and policy implications of converging new technologies integrated from the nanoscale. J. Nanoparticle Res. 7(2–3), 127–143.

Roco M.C., 2005b. International perspective on government nanotechnology funding in 2005. J. Nanoparticle Res. 7(6), 707–712.

22

Sayes C., J. Fortner, W. Guo, D. Lyon, A. Boyd, K. Ausman, et al., 2004. The differential cytotoxicity of water-soluble fullerenes. Am. Chem. Soc. 4, 1881–1887.

Sayes C., A. Gobin, K. Ausman, J. Mendez, J. West & V. Colvin, 2005. Nano-C60 cytotoxicity is due to lipid peroxidation. Biomaterials 26(36), 7587–7595.

Seifert, C., 15 July 2004. Industry Surveys. Insurance: Property-Casualty. Standard Poor's: NY.

Shvedova A., E. Kisin, R. Mercer, A. Murray, et al., 2005. Unusual inflammatory and fibrogenic pulmonary responses to single-walled carbon nanotubes in mice. Am. J. Physiol. Lung Cell. Mol. Physiol. 289(5), L698–L708.

Slaughter J., T. Lumley, L. Sheppard, J. Doenig & G. Shapiro, 2003. Effects of ambient air pollution on symptom severity and medication use in children with asthma. Allergy Asthma Immunol. 91(4), 346–353.

Swiss Re, 2004. Nanotechnology: Small Matter, many Unknowns. Swiss Re. Zurich, Switzerland.

The Royal Society, the Royal Academy of Engineering., 2004. Nanoscience and Nanotechnologies: Opportunities and Uncertainties. London, England: The Royal Society and the Royal Academy of Engineering.

Toxic Substances Control Act of 1976, § 3:15 U.S.C. § 2602.

U.S. Department of Agriculture Agricultural Research Service, 1997 [cited 5 July 2006]. EPA and Pesticide Registration Issues [Internet]. Available from: http://www.ars.usda.gov/is/np/mba/jan97/epa.htm.

U.S. Environmental Protection Agency, 2005. Nanotechnology White Paper – External Review Draft. Prepared by the Nanotechnology Workgroup, a group of EPA's Science Policy Council. Washington, DC.

Wang H., J. Wang, X. Deng, H. Sun, Z. Shi, Z. Gu, Y. Liu & Y. Zhao, 2004. Preparation and biodistribution of 125I-labeled water-soluble single-wall carbon nanotubes. J. Nanosci. Nanotechnol. 4(8), 1019–1024.

Warheit D., B. Laurence, K. Reed, D. Roach, G. Reynolds & T. Webb, 2004. Comparative pulmonary toxicity assessment of single-wall carbon nanotubes in rats. Toxicol. Sci. 77, 117–125.

Weiss R., 28 March 2005. Nanotech Is Booming Biggest in U.S., Report Says. Washington Post, p. A6.

Wellenius G., J. Schwartz & M. Mittleman, 2005. Air pollution and hospital admissions for ischemic and hemorrhagic stroke among medicare beneficiaries. Stroke 36(12), 2549–2553.

Wilhelm M. & B. Ritz, 2005. Local variation in CO and particulate air pollution and adverse birth outcomes in Los Angeles County, California, USA. Environ. Health Perspect. 113(9), 1212–1221.

Wood S., 2005. Knowing nano: understanding the risks of the science of the small Jupiter, Socially Responsible Investment and Governance Team. Kent, England.

Woodrow Wilson Center Project on Emerging Nanotechnologies, 2006 [cited 5 July 2006]. Nanotechnology. Environmental and Health Implications. A database of current research [Internet]. Washington, DC. Available from: www.nanotech-project.net/18.

Zheng M., A. Jagota, M. Strano, A. Santos, P. Barone, S. Chou, B. Diner, M. Dresselhaus, R. McLean, G. Onoa, G. Samsonidze, E. Semke, M. Usrey & D. Walls, 2003. Structure-based carbon nanotube sorting by sequence-dependent DNA assembly. Science 302(5650), 1545–1548.

Zhao X., A. Striolo & P. Cummings, 2005. C60 binds to and deforms nucleotides. Biophys. J. 89(6), 3856–3862.

Journal of Nanoparticle Research (2007) 9:23–38
DOI 10.1007/s11051-006-9159-5

© Springer 2006

Special focus: Nanoparticles and Occupational Health

Phospholipid lung surfactant and nanoparticle surface toxicity: Lessons from diesel soots and silicate dusts

William E. Wallace[1,2,]*, Michael J. Keane[1], David K. Murray[1], William P. Chisholm[1],
Andrew D. Maynard[3] and Tong-man Ong[1]
[1]*US National Institute for Occupational Safety and Health, 1095 Willowdale Road, Morgantown, WV,
26505, USA;* [2]*College of Engineering and Mineral Resources, West Virginia University, Morgantown, WV,
26506, USA;* [3]*Woodrow Wilson International Center for Scholars, Woodrow Wilson Plaza, 1300 Pennsylvania Avenue, N.W., Washington, DC, 2004-3027, USA;* *Author for correspondence (Tel.: + 304-285-6096;
E-mail: wwallace@cdc.gov)*

Received 24 July 2006; accepted in revised form 2 August 2006

Key words: nanoparticle, surfactant, particle surface, phospholipid, dipalmitoyl phosphatidyl choline,
surface analysis, toxicity, silicosis, genotoxicity, cytotoxicity, occupational health

Abstract

Because of their small size, the specific surface areas of nanoparticulate materials (NP), described as
particles having at least one dimension smaller than 100 nm, can be large compared with micrometer-sized
respirable particles. This high specific surface area or nanostructural surface properties may affect NP
toxicity in comparison with micrometer-sized respirable particles of the same overall composition.
Respirable particles depositing on the deep lung surfaces of the respiratory bronchioles or alveoli will
contact pulmonary surfactants in the surface hypophase. Diesel exhaust ultrafine particles and respirable
silicate micrometer-sized insoluble particles can adsorb components of that surfactant onto the particle
surfaces, conditioning the particles surfaces and affecting their *in vitro* expression of cytotoxicity or
genotoxicity. Those effects can be particle surface composition-specific. Effects of particle surface conditioning by a primary component of phospholipid pulmonary surfactant, diacyl phosphatidyl choline, are
reviewed for *in vitro* expression of genotoxicity by diesel exhaust particles and of cytotoxicity by respirable
quartz and aluminosilicate kaolin clay particles. Those effects suggest methods and cautions for assaying
and interpreting NP properties and biological activities.

Concerns for health hazard from nanoparticulate exposures

Research has demonstrated the importance of
parameters such as size and number in determining
the toxicity of insoluble particles with nanometer
dimensions, or nanoscale structures (Oberdörster
et al., 1995, 2004; Driscoll, 1996; Donaldson et al.,
2000; Oberdörster 2000; Tran et al., 2000).

Nanostructured materials including nanoparticles
(NP) are defined as having at least one dimension
smaller than 100 nm (Maynard & Kuempel, 2005).
There also is concern for surface property effects
on NP-induced toxicity or disease risk. This is due
to the large specific surface area, i.e., surface area
per unit mass, of NP associated with their small
size, and because surface area and surface properties can strongly affect the toxicity or disease risk

associated with respirable micrometer-sized particles. Therefore, while such possible effects are under investigation, NP as administered for cellular or animal model bioassay ideally should not be altered in size, morphology, aggregation and surface properties from their condition upon deposition in the lung after workplace or environmental inhalation exposure. As part of this, a critical concern is the conditioning of NP that will occur upon the initial deposition of particles upon the aqueous hypophase environmental interface of the lung, e.g., by adsorption of and dispersion in biomolecular components of lung surfactant or serum.

NP may differ on a mass basis from larger particles of the same composition for expression of toxicity and for biological transport and bioavailability because of higher specific surface area of NP. Materials deemed low in toxicity as larger particles may exhibit toxic effects as NP. Greater toxicity is reported for ultrafine carbon black, TiO_2 and latex particles compared to larger low-toxicity low-solubility particles of the same material (Donaldson et al., 2000); a tenfold increase in inflammation observed for the same mass of ultrafine versus fine particles was attributed to increased oxidative activities of the ultrafine particles. A set of dusts of low toxicity when in the micrometer size range including TiO_2, talc, carbon black, and photocopier toner, and particles with some toxicity including coal mine dust and diesel exhaust particulate material (DPM), were found to have comparable toxicity on a surface area basis for lung tumor induction in a rat model; and that toxicity increased strongly with increase in dose measured as surface area (Maynard & Kuempel, 2005). However, in some cases there is a strong mineral specific component of toxicity not resolved by surface area normalization of dose; for example, fine-sized crystalline silica, e.g., quartz dust, is much more active than TiO_2 for pulmonary inflammation in an animal model (Oberdörster et al., 1994). Differing degrees of inflammation and lung injury upon ultrafine NiO, Co_3O_4, TiO_2 and carbon black instillation in rat lung have been reported (Dick et al., 2003). Degree of lung injury was found to correlate to the particle's ability to generate surface free radicals and to cause oxidant damage. Surface area, chemical composition and surface reactivity were all deemed important factors in particle toxicity. Exacerbated pulmonary inflammation may be a means by which airborne pollutant matter (PM) exerts its toxicity (Tao et al., 2003). The smallest PM, below 2.5 μm, was most consistently associated with toxicity; the toxicity was attributed to oxidative stress caused by reactive oxygen species associated with metal, semi-quinone, lipopolysaccharide, or hydrocarbon constituents of ultrafine particles.

NP also may be able to cross the cell membrane and enter the bloodstream from the lungs (Ferin & Oberdörster, 1992; Oberdörster et al., 1992; Geiser et al., 2005). This general cell-penetrating ability is known, and even exploited, in the field of in vivo imaging. NP with special fluorescent, magnetic or optical properties such as "quantum dots" and magnetic resonance imaging contrast agents are functionalized with biocompatible coatings such as peptides, polysaccharides or other polymers and then directed within cells to permit selective signaling from specific cell components (Michalet et al., 2005; Sadeghiani et al., 2005). This ability to cross cell membranes has been pursued to provide functionalizing agents to transport peptides and DNA fragments into cells, e.g., through the endothelial tight-junction blood–brain barrier (Pantarotto et al., 2003, 2004a, b; Lu et al., 2005; Zhi et al., 2005). Such an uncommon effect is reported in studies of a variety of inorganic NP (Peters, et al., 2004). Cytotoxicity is a concern in new applications of NP, and safe exposure levels must be determined before these agents can be used in medical procedures. The majority of NP surveyed (TiO_2, SiO_2, and Co) were internalized into human epithelial cells, though most did not show cytotoxic effects. A pro-inflammatory stimulation and impairment of proliferative activity was observed for nano-Co and nano-SiO_2 particles, which was speculated to lead to a chronic inflammatory response and subsequent development of granulomas.

Carbonaceous materials represent a major class of NP, and a wide range of toxicity may result from variations in their shape, size, and complex chemical composition. The spherical nanoparticulate soot fullerene (C_{60}) was intentionally produced in 1985 by laser ablation of graphite targets. Limited toxicity studies of fullerene indicate this material was toxic to fish in aqueous systems, where fullerenes were found to pass the blood–brain barrier and cause brain damage (Oberdorster, 2004). The discovery of fullerenes has led to new categories of NP carbonaceous products with a variety of useful shapes, including multi-walled and single-walled carbon nanotubes (CNT). CNT may exhibit significant cell toxicity on the basis of the combined effects of quantum physical effects such as cell wall penetration and the toxicity observed for similar carbonaceous bulk materials. Other studies show evidence of toxic behavior, though the mechanisms involved are not described. CNT may inhibit HEK293 cell growth by inducing cell apoptosis and decreasing cellular adhesion ability (Cui et al., 2005). CNT adverse affects are reported for immortalized human epidermal keratinocytes (Shvedova et al., 2003). Oxidative stress and cellular toxicity was indicated by the formation of free radicals, the accumulation of peroxidative products, antioxidant depletion and loss of cell viability. Recent *in vitro* studies of NP have reported cytotoxic activities for C_{60} and multi-walled CNT, and also for Ag, TiO_2, Fe_2O_3, Al_2O_3, ZrO_2, Si_3N_4, carbon black, and MnO_2 NP (Bottini et al., 2005; Gurr et al., 2005; Hussain et al., 2005; Sayes et al., 2005; Soto et al., 2005).

In an animal model study, three single-walled carbon nanotube materials (SWCNT) containing different amounts of residual metals were intratracheally instilled into mice at three doses and histopathology performed at 7 and 90 d (Lam et al., 2004). All the SWCNT induced dose-dependent epithelioid granulomas and in some cases interstitial inflammation at 7 d, with progression at 90 d, demonstrating greater toxicity on a mass basis than a carbon black negative control and a quartz dust positive control. Somewhat different response was observed in a study using SWCNT instillation in a rat model at two doses, with bronchiolar lavage biomarker assay and tissue histopathology at 24 h, 1 week, 1 month, and 3 months (Warheit et al., 2004). There, SWCNT exposures resulted in transient inflammatory response and injury, with non-dose-dependent multi-focal granulomas that did not progress after 1 month, and lack of toxicity indicated by lung lavage and cell proliferation measures. To address the somewhat disparate results of these two studies, a complete evaluation was performed of the dose dependence and time course of pulmonary response of mice to single pharyngeal aspiration exposure of purified SWCNT at doses bracketing the equivalent of 20 workdays of exposure at the OSHA Permissible Exposure Limit (PEL) for

graphite particles (Shvedova et al., 2005). The study found acute inflammation and granulomatous response associated with dense SWCNT aggregates and, interestingly, early onset of progressive diffuse interstitial fibrosis and alveolar wall thickening associated with dispersed SWCNT distant from the aggregates. Protein, lactate dehydrogenase (LDH), and oxidative biomarkers were increased in bronchoalveolar lavage. Equal mass doses of ultrafine carbon black or fine crystalline silica dust caused weaker inflammation and damage, and no granulomas or cell wall thickening. *In vitro* macrophage exposures in the same study found TGF-beta 1 production and a weaker TNF-alpha and IL-1 beta production was stimulated by the SWCNT, but no stimulation of superoxide, NO, active engulfment, or apoptosis.

Genotoxicity of ultrafine DPM

DPM are anthropogenic, inadvertently generated organic NP materials. The National Institute for Occupational Safety and Health (NIOSH, 1988), the International Agency for Research on Cancer (IARC, 1989) and the US Environmental Protection Agency (USEPA, 2002) have declared diesel exhaust a potential or probable human carcinogen. DPM and carbonaceous materials from combustion processes are generally a complex mix of aromatic carbon graphitic sheets as a core, with reactive oxygen, nitrogen or sulfur heteroatomic functional groups, plus metals and organic species entrained during the synthesis or combustion process. Polycyclic aromatic hydrocarbon (PAH) compounds that are known carcinogens are contained in some of these products. Genotoxicity of carbonaceous material in NP may be attributed to PAH carcinogenicity, oxidation from reactive oxygen species (ROS) activity, nitrate generation, or transition metal chemistry. In vivo animal studies of DPM have confirmed lung tumors in the rat from long-term inhalation exposures (Heinrich et al., 1986; Mauderly et al., 1987). However, the genotoxic role of DPM in inhalation tumorogenicity studies has been questioned after comparable tests of non-genotoxic carbon black resulted in tumorogenesis in the same animal models (Heinrich et al., 1994; Nikula et al., 1995). DPM are insoluble in water, and typically are prepared for chemical study or *in vitro* bioassay by solution in organic solvents and fractionation by chromatographic techniques. DPM typically contain a number of proven toxic species; polar hetero- and polycyclic aromatics, radical species, entrained metal species, and entrained organic solvent species, with the composition varying with engine performance characteristics, fuel types, lube oil types, and extraction solvents (Morimoto, 1986). These species have been fractionated and measured in DPM using bulk mass, particle size, as well as advanced chromatographic and mass spectroscopic techniques.

In vitro genotoxic activities elicited by organic solvent extracts, e.g., acetone or dichloromethane solvent extracts of some filter-collected DPM, are well-reported (IARC, 1989). The use of extraction solvents has led to discrepancies in assigning genotoxic effects to chemical characteristics (Hayakawa et al., 1997; Soontjens et al., 1997; Saxena et al., 2003; Siegel et al., 2004). Apparent false negatives or positives have been attributed to solvent/adsorbed organic dispersion, solvency or matrix effects. However, the *in vitro* mutagenicity of solvent extract of DPM can vary systematically with operating conditions for a given engine, e.g., with engine speed and torque (McMillian et al., 2002).

The lack of water-solubility of the polycyclic organics from DPM raised the question of their biological availability for genotoxic activity under conditions of particle deposition in the lung. It was recognized that the conventional procedure of testing organic solvent extracts of DPM did not necessarily provide a physiologically reasonable model of genotoxicant biological availability as manifest by intact particles deposited in the lung. Therefore, research examined as a medium for *in vitro* cellular challenge, the extract of DPM by primary components of the surfactant hypophase layer that coats the deep lung respiratory bronchioles and alveoli, frequently using diacyl phosphatidyl cholines dispersed into physiological saline.

Surrogate lung surfactant

The environmental interface of the deep lung respiratory bronchioles and alveoli for initial contact with inhaled particles is the surfactant-coated and laden hypophase (reviewed in Bourbon (1991)). Lipids and lipoprotein surfactants are synthesized and secreted onto the wet surface of the deep lung by alveolar type II cells. Additional biochemical ingredients, e.g., components of mucus, are found in the hypophase of the upper airways as part of the mucociliary system for lung clearance. Phospholipids are a major component of lung surfactant. By themselves they can reproduce physiologically important surface-tension properties of the pulmonary alveolar hypophase surface (Scarpelli, 1968). Research modeling lung surface tension phenomena often has used a major phospholipid component of pulmonary surfactant, dipalmitoyl phosphatidyl choline (DPPC) or some other diacyl phosphatidylcholines, dispersed into physiologic saline, as a simple model of lung surfactant. This surfactant has been used to model the possibility of lung surfactant extraction of genotoxicants from DPM. Experiments incorporated DPPC dispersed into physiological saline as a solvent to attempt to extract filter-collected DPM. However, this produced an extract with little or no genotoxic activity (Brooks et al., 1981; King et al., 1981; McClellan et al., 1982; Wallace et al., 1987), even when the organic solvent extract of a parallel DPM sample expressed significant activity. Instead, it was found that some DPM can express genotoxic activity, e.g., DNA or chromosomal damage *in vitro*, as a dispersion in this surfactant. DPM was tested as a direct mixture into such surfactant, without subsequent filtering; that is, the DPM was tested as a non-dissolved but surfactant-dispersed particulate in surfactant mixture (Wallace et al., 1987, 1990a; Keane et al., 1991; Gu et al., 1992, 1994, 2005). DPM genotoxic activities are expressed when the DPM is dispersed into DPPC surfactant; and those activities are associated with the non-dissolved particulate phase material. DPPC dispersion does not extract genotoxicants from the DPM particles; rather, the phospholipid coats and "solubilizes" (not "dissolves") the DPM, providing a hydrophilic coating and permitting the dispersion of the surfactant-coated DPM as particles in aqueous media.

To first order, this dispersion of collected DPM into the principal component of lung surfactant should adulterate the collected DPM approximately to the degree that the particles would be conditioned upon deposition in the deep lung, causing particle agglomeration or disassociation and particle surface conditioning to approximately the same degree as would occur in the lung alveolar hypophase. That is, assaying DPM dispersed (mixed) into lavaged or synthetic models of lung alveolar hypophase surfactant avoids to first order the non-physiologic dissolution of particles and associated destruction of particulate properties, as would occur in organic solvent extraction of collected DPM. This provides a physiologically reasonable representation *in vitro* of toxicant bioavailability for particles depositing in the lung.

After filter-collected DPM is mixed into a dispersion of DPPC surfactant in saline, the then wettable-surfaced nano-particles can challenge cells effectively to express genotoxic activities for mammalian cell DNA and clastogenic damage, as well as for bacterial cell mutagenicity. This provides a physiologically reasonable method for handling and delivery of DPM for toxicological assays which might be applicable to other insoluble hydrophobic NP materials. And the phenomenon of surfactant-dispersed DPM expression of genotoxic activities suggests a first mechanistic step by which cells can be effectively exposed to insoluble NP genotoxicants following inhalation and deposition on the deep lung pulmonary respiratory bronchioles or alveoli.

The basis for the surfactant activity of DPPC is that the molecular structure contains a hydrophilic end and a hydrophobic end (Figure 1). The former consists of a trimethyl ammonium cationic group bound through a two-carbon chain to an acidic phosphate, forming a zwitterionic dipole and providing a hydrophilic moiety as one end of the molecule. The phosphate is esterified to the first carbon of a glycerol which is esterified at the other two carbons to two long chain fatty acid residues,

palmitate in the case of DPPC. These provide two hydrophobic, lipophilic long tails to the molecule. When dispersed into aqueous media, the phospholipid molecules aggregate into multi-molecular structures such that the hydropilic zwitterionic head groups of the molecules are oriented to face into the surrounding water molecules, while the hydrophobic fatty acid tails cluster amongst themselves, minimizing contact with water or the hydrophilic heads of other phospholipids. This gives rise to spherical or lamellar structures made up of a bilayer of surfactant molecules. The zwitterionic head groups are on the outer surfaces of the bilayer, with the lipid tails "sandwiched" between. This structure is the general basis for the bilayer phospholipid membrane underlying cell membrane structure.

DPPC can be dispersed into these liposomal and lamellar bilayer structures by ultrasonication into saline, forming a pale milky stable dispersion. When collected DPM is mixed into this DPPC dispersion the DPM soot particle agglomerates are observed to disperse. The DPM is "solubilized", that is, dispersed as small particles rather than dissolved into the DPPC dispersion. In this DPM-in-DPPC dispersion, the long chain lipophilic/hydrophobic tails of the DPPC molecule associate with the organic DPM particle surfaces, while the zwitterionic hydrophilic trimethyl ammonium and phosphate head of the molecule orients outward to face the surrounding aqueous medium. A simplified picture is that of a DPM particle as a tar "pin-cushion" covered by DPPC soap molecules with their tails stuck as the shaft of pins to the pin-cushion DPM particle and their heads outward, providing a hydrophilic outer coating, in-turn permit-

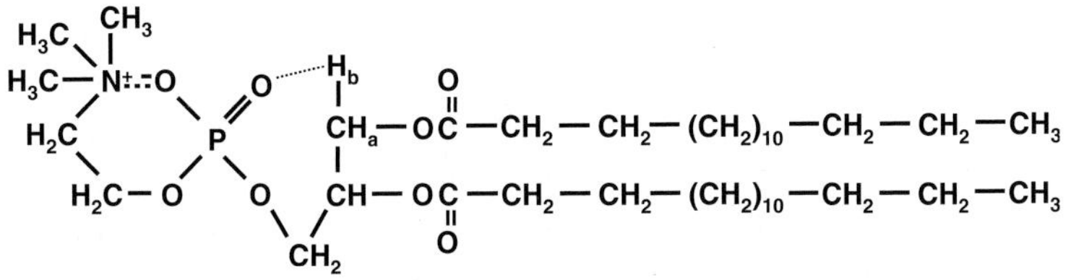

Figure 1. DPPC surfactant structure: Palmitate residues associate with DPM hydrocarbon; The zwitterionic trimethyl ammonium – acidic phosphate end of the molecule is hydrophilic, the two fatty acid tails esterified to the glycerol are hydrophobic and lipophilic, leading the molecules to aggregate in structures in aqueous media with the hydrophilic moieties oriented toward the water. The hydrophobic end of the molecule will associate with those tails of other lipid molecules or with the surface of hydrophobic particles, e.g., diesel exhaust particles, providing a "wettable" surface on a surfactant-coated DPM particle.

ting the structure to act as a water-wet but non-dissolved NP which disperses in water.

In vitro genotoxicity assays of DPM dispersed in surfactants

Comparisons of *in vitro* genotoxicites have been made between surfactant-dispersed and solvent-extracted DPM (Wallace et al., 1987, 1990a; Keane et al., 1991; Gu et al., 1992, 1994, 2005).

The DPM tested was filter-collected and graciously supplied by the Lovelace-Inhalation Toxicology Research Institute from a 1980 GM 5.7 liter V8 engine run on the Federal Test Procedure Urban Driving Cycle. For organic solvent extracted sample, DPM was dissolved into dichloromethane (DCM) and evaporatively exchanged into dimethylsulfoxide (DMSO) at 1 mg DPM/ml DMSO. For the surfactant dispersion sample, the surfactant was prepared by ultrasonically dispersing DPPC into physiological saline (PSS) at 2.5 mg DPPC/1 ml PSS; then DPM was mixed into that dispersion at 1 mg DPM/ 2.5 mg DPPC / 1 ml PSS. The tested materials were (a) the total preparations, in DMSO and in DPPC/PSS; (b) supernatants from centrifugation and filtration of the total preparations; and/or (c) sediments from centrifugation of the total prepa-

rations, i.e., the non-dissolved particulate phase material.

The Ames *Salmonella typhimurium* TA98 was used for the mutagenicity assay (Ames et al., 1975). As shown in Figure 2 (Keane et al., 1991), the total dispersion samples, both solvent and surfactant total preparations, were comparably mutagenic. The supernatant fraction (extracted material) of the solvent preparation was positive and of the surfactant preparation supernatant was negative; i.e., no active mutagenic material extracted from DPM by surfactant. The sediment fraction (non-dissolved particulate material) for the solvent preparation was negative, i.e., the carbonaceous residue of solvent extracted DPM was not mutagenic, while the surfactant preparation was positive, i.e., the particulate matter which is not dissolved by surfactant is, nevertheless, positive for mutagenic activity as a particulate dispersion in surfactant. Other tests of the surfactant preparation supernatant fraction using only centrifugation without subsequent filtration resulted in some activity in the surfactant preparation supernatant. This was interpreted as due to ultrafine particles of surfactant-dispersed DPM that were fully removed by filtration but not fully removed by centrifugation.

Chinese hamster pulmonary fibroblast-derived cell line (V79) was used to test for the induction of

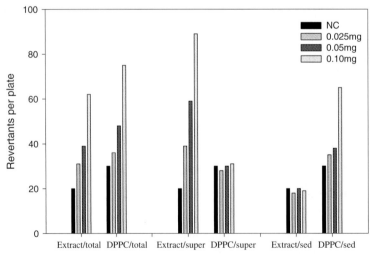

Figure 2. Mutagenic activity (TA98 *Salmonella typhimurium* TA98) versus DPM concentration: Mutagenic activity as number of revertant colonies is shown (y-axis) versus DPM concentration as solvent extract or DPPC surfactant dispersion. Activities are shown for the total solvent or surfactant preparation, for the filtered supernatant, and for the non-dissolved sediment. From: Wallace et al. (1987).

sister chromatid exchange (SCE) (Perry & Wolff, 1974), unscheduled DNA synthesis (UDS) (Mitchell et al., 1983) and chromosomal aberration (Preston et al., 1981). Results of the SCE assay as shown in Figure 3 (Keane et al., 1991) were similar to those of the Ames mutagenicity assay. Both organic extracted and surfactant-dispersed DPM materials induced SCE in V79 cells. After separation of the samples into supernatant and sediment fractions, the activity of both DPM preparations was found to reside in the supernatant fraction of the solvent-extracted samples, and in the sedimented fraction for surfactant dispersed samples.

Results of UDS assay are shown in Figure 4 (Gu et al., 1994). Both dispersions of DPM in surfactant and DMSO induced UDS in a concentration related manner. Again, induction of UDS was also found in the supernatant fraction of the DMSO-dispersed sample and in the sedimented fraction of the surfactant-dispersed sample. Chromosomal aberration (CA) studies found that surfactant-dispersed DPM was active for induction of CAs, generally increasing with DPM concentration (Gu et al., 2005).

Assay of induction of micronuclei (Lansne et al., 1984) using V79 cells and in Chinese hamster ovary-derived cells (CHO) was measured for DPM solvent extract and/or surfactant dispersion supernatant and particulate phase materials (Gu

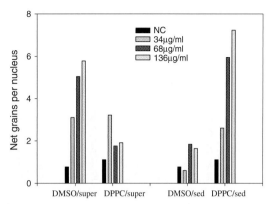

Figure 4. UDS (V79 mammalian cell) versus DPM concentration: UDS is represented as net autoradiographic grains/nucleus (*y*-axis) versus DPM concentration as solvent extract or DPPC surfactant dispersion; for total preparation, filtered supernatant, and sediment. From: Gu et al. (1994).

et al., 1992). The solvent supernatant (extract) of total samples after centrifugation and filtration, and the surfactant sediments (particulate material) from total sample centrifugation were active for micronucleus induction in CHO cells: the solvent extract was active in V79 cells, but the surfactant sediment was only marginally active in V79 cells (Figure 5).

The results from these studies with different genetic endpoints in bacteria and in mammalian

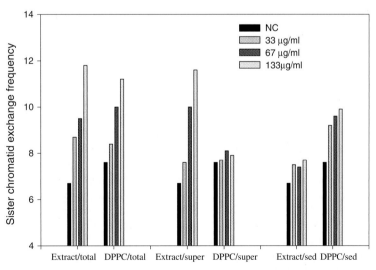

Figure 3. SCE (V79 mammalian cell) versus DPM concentration: Number of SCE/cell is shown (*y*-axis) versus DPM concentration as solvent extract or as DPPC surfactant dispersion, for the total preparation, filtered supernatant, and sediment. From: Keane et al. (1991).

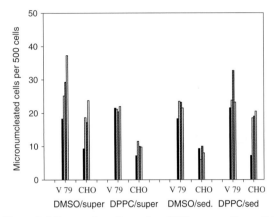

Figure 5. Micronucleus formation (V79 mammalian cells) versus DPM concentration: Micronucleus frequency per 500 cells (*y*-axis) is shown versus DPM concentration as solvent or surfactant filtered supernatant or sediment. From: Gu et al. (1992).

cells are consistent and show that DPM genotoxic activities are expressed when the DPM is dispersed into DPPC surfactant; and those activities are associated with the non-dissolved particulate phase material. DPPC dispersion does not extract genotoxicants from the DPM particles; rather, the phospholipid coats and "solubilizes" (not "dissolves") the DPM, providing a hydrophilic coating and permitting the dispersion of the surfactant-coated DPM in aqueous media. These findings indicate that genotoxic activity and potential carcinogenicity associated with DPM inhaled into the lung may be made bioavailable by virtue of the solubilization/dispersion properties of pulmonary surfactant components.

Fine respirable mineral particle surface nanostructure and disease risk

Mineral particle composition can strongly affect toxic activity and disease risk associated with respirable dust exposures. One of the most studied mineral dusts is respirable crystalline silica, e.g., quartz, a known potent agent for pulmonary fibrosis (Green & Vallyathan, 1995), and evaluated by the International Agency for Research on Cancer (IARC) to be a human carcinogen under some exposure conditions. Quartz dust also expresses some much higher in vivo toxicities in animal models than a number of other mineral dusts and other respirable materials when the dose

metric is surface area. However, even beyond this mineral specificity, quartz dust-induced human disease risk is affected significantly by sub-micrometer scale surface coatings by heteroatomic materials. Seemingly anomalous differences in lung fibrosis disease risk from silica dust exposures in mixed dust composition atmospheres have been observed, e.g., in coal workers' pneumoconiosis (Attfield & Morring, 1992), and these anomalies have been associated with the existence of sub-micrometer thick mineral coatings or "occlusion" of silica particles by aluminosilicate. In animal model experiments, some workplace silica dusts were found to be much diminished in fibrogenic activity compared to the dusts after acid-etching, suggesting a prophylactic surface coating on the silica particles, e.g., of aluminosilicate clay (Le-Bouffant et al., 1982). Spectroscopic analyses have shown such sub-micrometer thick aluminosilicate coating or "occlusion" of respirable silica particles, detected by contrasting scanning electron microscopy - energy dispersive X-ray analysis of individual silica particle composition at high and at low electron beam energies to probe particle composition with depth (Wallace et al., 1990b, 1992, 1994; Wallace & Keane, 1993; Hnizdo & Wallace, 2002). Such sub-micrometer coatings or occlusion of quartz particle surfaces by clay have been associated with the epidemiological "coal rank anomaly" in the prevalence of coal workers' pneumoconiosis (Walton et al., 1971; Robock & Klosterkotter, 1973; Kreigseis & Scharmann, 1982; Attfield & Morring, 1992). Research indicated that the fraction of silica particles surface occluded by aluminosilicate decreased with increasing coal rank, with a consequent greater fraction of silica particles with "biologically available" surface for dusts from mines of higher rank coals (Harrison et al., 1997). The equivalent was seen in a study of anomalous differences in silicosis risk between Chinese metal mine workers and pottery workers quantified in a silicosis medical registry of some 20,000 workers (Chen et al., 2005). Normalizing the workers' cumulative respirable silica dust exposures to cumulative respirable "surface-available" silica dust, i.e., the fraction of respirable silica dust particles not surface occluded by clay aluminosilicate, resolved much of the difference in risk (Harrison et al., 2005).

Nano-scale surface composition and structure were found to be important for dust toxicity and

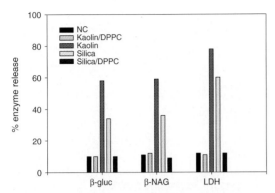

Figure 6. Quartz and kaolin expression of in vitro cyto-toxicities: Cytotoxicities are shown (*y*-axis) as % release of cellular cytosolic enzyme LDH , and cellular lysosomal enzymes beta-glucuronidase, and beta-*N*-acetyl glucosa-minidase. Native quartz and kaolin dusts are comparably cytotoxic on a surface area basis; and both have their tox-icities fully suppressed immediately after incubation with DPPC surfactant. From: Wallace et al. (1988).

ated in a new hard metal production facility where sentinel adverse health effects had been observed. The structure of those coatings, i.e., thin Co in contact with the underlying WC particle surface, resulted in heightened levels of catalytically gen-erated reactive oxygen species in aqueous media; this led to associated exacerbation of *in vitro* tox-icities in comparison to dusts from a conventional fabrication process (Keane et al., 2002a, b).

Surfactant effects on mineral particle toxicity

hazard in a "hard metal" manufacturing work-place, e.g., producing tungsten carbide (WC) grit cemented together by cobalt metal to form hard sharp edges for cutting tools. Surface analyses by scanning Auger spectroscopy in combination with scanning electron microscopy – X-ray spectros-copy (Stephens et al., 1998) found nano-meter thin cobalt coating on respirable WC particles gener-

Exposure to respirable aluminosilicate kaolin clay dust does not present the high risk of lung fibrosis presented by quartz dust; and aluminosilicate surface coatings on quartz particles are associated with diminished silicosis risk. Nevertheless, kaolin dust expresses *in vitro* cytotoxic activities compa-rable to that of quartz dust on a surface area basis, as measured by mammalian cell release of LDH and lysosomal enzymes or by erythrocyte mem-branolysis (Figure 6) (Vallyathan et al., 1988; Wallace et al., 1989, 1992). Thus, aluminosilicate dust false positives prevent the effective use of short term *in vitro* assays to predict disease hazard. This suggests that, similarly, *in vitro* cytotoxicity assays of NP may not be directly predictive of disease risk. Again, respirable particle condition-

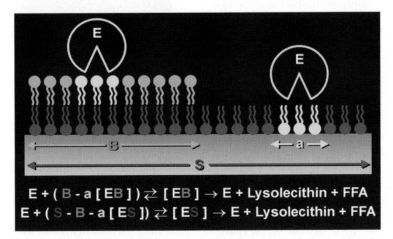

Figure 7. Bi-layer model of phospholipase digestion of phospholipid surfactant adsorbed on a mineral particle surface: An inner layer of DPPC molecules (S) adsorb to hydrophilic mineral dust surface; this is backed by a DPPC layer oriented in the reverse direction (B). Phospholipase A₂ enzyme (E) digests the outer layer with a rapid and non-mineral-specific rate, producing lyso (monoacyl) phosphatidylcholine and free palmitic fatty acid. Enzyme access to the inner layer is partially hindered by the outer layer of DPPC (B), by the presence of adjacent enzymes (a), and by mineral-specific surface functional groups interacting with adsorbed DPPC, e.g., DPPC phosphate binding to clay surface aluminol, to result in different conformations of the adsorbed DPPC with consequent differences in rates of removal and restoration of particle toxicity. From: Wallace et al. (1994b).

ing by pulmonary surfactants will occur upon deposition of mineral particles in the deep lung (Emerson & Davis, 1983). DPPC is adsorbed from dispersion in physiological saline by quartz (Jaurand et al., 1979) or by kaolin (Wallace et al., 1975, 1985) and suppresses the dusts' otherwise prompt *in vitro* cytotoxicity.

However, *in vitro* assays of mineral dusts including modeling of surfactant conditioning do not provide unambiguous prediction of disease risk; e.g., surfactant treatment suppresses the toxicity of quartz for a period of time, presenting an apparent false negative assay result. Thus the next in vivo event must be considered: the cellular uptake and enzymatic digestion of the particles with possible hydrolysis and removal of prophylactic surfactant from the particles and consequent restoration of toxicity over time. In acellular and *in vitro* cellular studies, phospholipase A2 (PLA2) enzyme digests DPPC from the dusts, with restoration of cytotoxic activity. Kinetics of the process is well-modeled as a two-exponential function process with an outer layer of DPPC molecules rapidly hydrolyzed at the glycerol ester linkages by the lipase, while the inner DPPC layer in direct contact with the mineral surface is digested more slowly and with mineral specificity (Figure 7). For extracellular PLA2 acting at neutral pH, the inner layer DPPC is digestively removed much more slowly from kaolin than from quartz (Wallace et al., 1988, 1992; Hill et al., 1995; Liu et al., 1996, 1998; Das et al., 2000; Keane et al., 1990, 2005). This is associated with mineral-specific conformations of adsorbed DPPC (Murray et al., 2005) in which interaction of the DPPC phosphate with kaolin surface aluminol groups confers an added steric hindrance to the PLA2 hydrolytic activity on the carbonyl ester adjacent to the phosphate. This difference in rates for surfactant removal is seen for quartz versus kaolin particles phagocytosed by pulmonary macrophage- derived cells e.g., under conditions in the acidic pH phagolysosome (Figure 8); but kaolin also is stripped of surfactant and restored to activity, albeit after a longer assay incubation of several days (Keane & Wallace, 2005). This suggests that a predictive assay would require the use of cell systems modeling the neutral pH phagolysosomal systems of interstitial cells (Adamson et al., 1989; Johnson & Maples, 1994).

DPPC is a limited model of lung surfactant, and of other biological molecules that may be found in

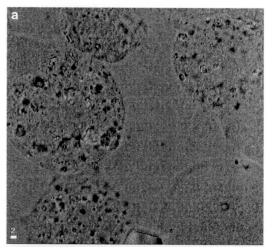

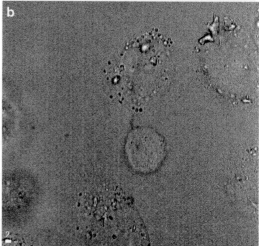

Figure 8. Removal of fluorescent-labeled phospholipids on kaolin (a) or quartz (b): Fluorescence microscopy images phospholipid remaining on phagocytosed quartz or kaolin particles at 8 Days after challenge. A green fluorescent boron-labeled analogue of diacyl phosphatidylcholine is shown retained on kaolin particles in cells (a), but lost from silica particles (b) with associated expression of cell toxicity.

the lung alveolar hypophase. Mineral specificity of prophylaxis and rate of restoration of expression of toxicity may be affected additionally by other lipids or lipoproteins. Two sizes of industrial ultrafine carbon blacks were found to adsorb significant DPPC and Surfactant Protein-D from aqueous dispersion, affecting particle agglomeration and precipitation; while little effect was noted for their incubation with fibrinogen or albumin (Kendall et al., 2004). Lipoprotein fractions of cell test system media serum can reduce the expression of crystalline silica cytotoxicity (Kozin & McCarty, 1977; Barrett et al., 1999a, b), with re-activation following trypsin digestion (Fenoglio et al., 2005). Quartz and kaolin dust prompt *in vitro* induction of LDH release from macrophage is suppressed in 10% fetal bovine serum (FBS) medium; however, quartz but not kaolin activity was restored at 6 h (Gao et al., 2000, 2001, 2002). This indicates that short-term *in vitro* assay results can be affected by assay system nutrients that are not necessarily representative of in vivo pulmonary hypophase exposures. However, in vivo acute inflammatory reactions are accompanied by increased permeability of the microvasculature with transudation of plasma protein, including albumin, (Slauson & Cooper, 1990; Driscoll, 1994, 1996) into the lung alveoli. Thus, following non-mineral specific response of alveolar macrophages to quartz or kaolin, the subsequent inflammatory response may provide a second-tier and mineral specific prophylaxis to deposited particles, conditioning their subsequent interstitial interactions.

Lessons for NP studies

Diesel exhaust nanoparticulate material and respirable micrometer-sized mineral dust expression of *in vitro* cytotoxicity or genotoxicity can be strongly affected by particle surface conditioning by a phospholipid component of lung surfactant, modeling an initial in vivo phenomenon not usually considered for assays of respirable particle toxicities. Detailed surface structural features not considered in conventional industrial hygiene characterizations can significantly affect disease risk associated with respirable mineral dust exposures. NP have high specific surface areas. Therefore, particle surface composition and its pulmonary surfactant conditioning should be considered in the design and interpretation of *in vitro* cytotoxicity or genotoxicity assays of NP, and for in vivo assays which might involve disruption of the pulmonary hypophase in the lung of the animal model, e.g., in some localized regions during instillation challenge.

Experience with diesel NP suggests that surfactant conditioning may provide a physiologically plausible method for assay of other carbonaceous or organic NP material potential for genotoxic activity under conditions modeling particle deposition in the deep lung. Research has shown that DPPC, a major phospholipid component of lung surfactant, does not extract genotoxicant compounds from diesel exhaust NP; rather, the surfactant coats the particles and those non-dissolved but surface-conditioned particles are able to express genotoxic activity. This has been demonstrated for bacterial mutagenicity and for mammalian cell DNA and clastogenic damage in a number of specific assay endpoints, e.g., SCE, UDS, chromosomal aberrations, micronucleus induction, and single cell gel electrophoresis-visualized DNA strand breaks. The amount of surfactant used in pre-conditioning the NP should be in excess of that required for bilayer surface coverage of the NP material. Measures of *in vitro* membranolytic activity suppression versus amount of surfactant adsorption on respirable quartz and kaolin dusts indicate this is approximately 5 mg DPPC per square meter of dust surface, the dust surface independently measured by nitrogen gas adsorption isotherms (Wallace et al., 1992). However, results indicate that additional DPPC will adsorb to several additional molecular layers on the dusts. In the case of diesel exhaust NP there is evidence that surfactant amounts on the order of equal mass of DPPC per mass of DPM are required for maximum expression of toxicity (Wallace et al., 1987). This may reflect extremely high specific surface areas of some DPM for surfactant adsorption, or a preparative effect of surfactant dispersion concentration on the agglomerative state of sampled material.

Such surfactant surface conditioning of inorganic respirable particles does not immediately provide a similarly direct and effective method for assay of non-genotoxic disease hazard; and these caveats may apply also to NP. In the absence of surfactant surface conditioning, a false positive assessment can result from the innate surface toxicity of some minerals; that is, toxic surface interactions observed *in vitro* are not manifest as in vivo pathology due to physiologic prophylaxis systems in the lung. For instance, assay of alumina or aluminosilicate NP in the absence of physiological surfactant or serum pre-conditioning risks the false positive results (for prediction of fibrosis) predicted by *in vitro* membranolysis and cytokine release assays of micrometer-sized respirable kaolin dust. The converse problem also is possible: short term assays can give a false negative result because of physiological but transient prophylaxis by surfactant conditioning of toxic dusts, whose true hazard might be revealed by longer assay challenge times, e.g., even quartz or other hazardous dust expression of toxic activity may be delayed several days in vivo and *in vitro* by surfactant surface conditioning. And the use of supplemented serum to nourish cells in longer-term *in vitro* assays can risk a false negative assay result from prophylaxis effects of nutrient components not representative of the in vivo milieu. There is a general caveat to the use of DPPC or any simplified model of lung surfactant for such surface conditioning for bioassay: lung surfactant is a complex lipoprotein mixture and the composition of potentially prophylactic biomolecules in the alveolar hypophase can change with exposure, e.g., with inflammation-associated transudation of albumin or other serum constituents into the lung. Thus the adsorption and possible resultant prophylactic effects of these different biomolecular components of the lung hypophase may differ, perhaps with different particle surface specificities.

Research on surfactant and serum interactions with respirable particle surfaces has indicated profound effects on the expression of toxicity suggesting that interactions of respired NP with biological molecular constituents of the hypophase liquid lining of the lung should be considered in the preparation and interpretation of bioassays of potential NP respiratory hazard.

Disclaimer: The findings and conclusions in this report are those of the authors and do not necessarily represent the views of the National Institute for Occupational Safety and Health.

Acknowledgment

Research on surfactant-dispersed diesel exhaust particulate genotoxicity was supported in part by the US Department of Energy – FreedomCar and Vehicle Technologies Activity.

References

Adamson I.Y.R., Letourneau H.L. & D.H. Bowden, 1989. Enhanced macrophage-fibroblast interactions in the pulmonary interstitium increases after silica injection to monocyte-depleted mice. Am. J. Physiol. 134, 411–417.

Ames B., J. McCann & E. Yamasaki, 1975. Methods for detecting carcinogens and mutagens with the Salmonella/mammalian microsome mutagenicity test. Mutat. Res. 31, 347–364.

Attfield M.D. & K. Morring, 1992. An investigation into the relationship between coal workers pneumoconiosis and dust exposure in US coal mines. Am. Indust. Hyg. Assoc. J. 53(8), 486–492.

Barrett E.G., C. Johnston, G. Oberdorster & J.N. Finkelstein, 1999a. Silica binds serum proteins resulting in a shift of the dose-response for silica-induced chemokine expression in an alveolar type II cell line. Toxicol. Appl. Pharmacol. 161, 111–122.

Barrett E.G., C. Johnston, G. Oberdorster & J.N. Finkelstein, 1999b. Antioxidant treatment attenuates cytokine and chemokine levels in murine macrophages following silica exposure. Toxico. App. Pharmaco. 158, 211–220.

Bottini M., S. Bruckner, K. Nika, N. Bottini, S. Bellucci, A. Magrini, A. Bergamaschi & T. Mustelin, 2005. Multi-walled carbon nanotubes induce T lymphocyte apoptosis. Toxicol. Lett. 160, 121–126.

Bourbon J.R., ed., 1991. Pulmonary Surfactant: Biochemical, Functional, Regulatory, and Clinical Concepts. CRC Press, Inc., 438 pp.

Brooks A.L., R.K. Wolff, R.E. Royer, C.R. Clark, A. Sanchez & R.O. McClellan, 1981. Deposition and biological availability of diesel particles and their associated mutagenic chemicals. Environ. Int. 5, 263–268.

Chen W., E. Hnizdo, J.-Q. Chen, M.D. Attfield, P. Gao, F. Hearl, J. Lu & W.E. Wallace, 2005. Risk of silicosis in cohorts of Chinese tin and tungsten miners and pottery workers (I): An epidemiological study. Am. J. Indust. Med. 48, 1–9.

Cui D, F. Tian, C.D. Ozkan, M. Wang & H. Gao, 2005. Effect of single wall carbon nanotubes on human HEK293 cells. Toxicol. Lett. 155(1), 73–85.

Das A., E. Cilento, M.J. Keane & W.E. Wallace, 2000. Intracellular surfactant removal from phagocytized minerals: Development of a fluorescent method using a BODIPY labelled phospholipid. Inhal. Toxicol. 12, 765–781.

Dick C.A.J., D.M. Brown, K. Donaldson & V. Stone, 2003. The role of free radicals in the toxic and inflammatory effects of four different ultrafine particle types. Inhal. Toxicol. 15, 39–52.

Donaldson K., V. Stone, P.S. Gilmore, D.M. Brown & W. MacNee, 2000. Ultrafine particles: Mechanisms of lung injury. Phil. Trans. R. Soc. Lond. A 358, 2741–2749.

Driscoll K.E., 1994. Macrophage inflammatory proteins: Biology and role in pulmonary inflammation. Exp. Lung. Res. 20, 473–490.

Driscoll K.E., 1996. Role of Inflammation in the Development of Rat Lung Tumors in Response to Chronic Particle Exposure. In: Mauderly J.L., & McCunney R.J. ed. Particle Overload in the Rat Lung and Lung Cancer: Implications for Human Risk Assessment. Taylor 7 Francis, Philadelphia, pp. 139–153.

Emerson R.J. & G.S. Davis, 1983. Effect of alveolar lining material-coated silica on rat alveolar macrophahges. Environ. Health Persp. 51, 81–84.

Fenoglio F., F. Gillio, M. Ghiazza & B. Fubini, 2005. Modulation of free radical generation at the surface of quartz by albumin adsorption/ digestion. [Abstract] In Mechanisms of action of inhaled fibres, particles, and nanoparticles in lung and cardiovascular disease. Research Triangle Park, NC; October 26–28.

Ferin J. & G. Oberdorster, 1992. Translocation of particles from pulmonary alveoli into the interstitium. J. Aerosol Med. Deposit. Clear. Effects Lung 5(3), 179–187.

Gao N., M.J. Keane, T. Ong & W.E. Wallace, 2000. Effects of simulated pulmonary surfactant on the cytotoxicity and DNA-damaging activity of respirable quartz and kaolin. J. Toxicol. Environ. Health 60(A), 153–167.

Gao N., M.J. Keane, T. Ong, J. Ye, W.E. Miller & W.E. Wallace, 2001. Effects of phospholipid surfactant on apoptosis induction by respirable quartz and kaolin in NR 8383 rat pulmonary macrophages. Toxicol. Appl. Pharmacol. 175, 217–225.

Gao N., M.J. Keane, T. Ong, J. Martin, W. Miller & W.E. Wallace, 2002. In: Respirable Quartz and Kaolin Aluminosilicate Expression of In vitro Cytotoxicity and Apoptosis in the Presence of Surfactant or Serum: Caveats to Bioassay Interpretation; Proceedings of Inhaled Particles IX. Cambridge UK. Ann. Occupat. Hyg. 46(S1), 50–52.

Geiser M., B. Rothen-Rutishauser, N. Kapp, S. Schurch, W. Kreyling, H. Schulz, M. Semmler, V.I. Hof, J. Heyder & P. Gehr, 2005. Ultrafine particles cross cellular membranes by nonphagocytic mechanisms in lungs and in cultured cells. Environm. Health Persp. 113, 1555–1560.

Green H.Y.F. & V. Vallyathan, 1995. Pathologic Responses to Inhaled Silica. In: Castranova V., Vallyathan V., & Wallace W. ed. Silica and Silica-Induced Lung Disease: Current Concepts. CRC Press, Boca Raton FL, pp. 163–185.

Gu Z.W., B.Z. Zhong, B. Nath, W.Z. Whong, W.E. Wallace & T. Ong, 1992. Micronucleus induction and phagocytosis in mammalian cells treated with diesel emission particles. Mutat. Res. 279, 55–60.

Gu Z.W., B.Z. Zhong, M.J. Keane, W.Z. Whong, W.E. Wallace & T. Ong, 1994. Induction of unscheduled DNA synthesis in V79 cells by diesel emission particles dispersed in simulated pulmonary surfactant. Ann. Occupat. Hyg. 38(1), 345–349.

Gu Z.W., M.J. Keane, T. Ong & W.E. Wallace, 2005. Diesel exhaust particulate matter dispersed in a phospholipid surfactant induces chromosomal aberrations and micronuclei but not 6–thioguanine-resistant gene mutation in V79 cells in vitro. J. Toxicol. Environ. Health 68(A), 431–444.

Gurr J.R., A.S. Wang, C.H. Chen & K.Y. Jan, 2005. Ultrafine titanium dioxide particles in the absence of photoactivation can induce oxidative damaeg to human bronchial epithelial cell. Toxicology 213, 66–73.

Harrison J., P. Brower, M.D. Attfield, C.B. Doak, M.J. Keane & W.E. Wallace, 1997. Surface composition of respirable silica particles in a set of US anthracite and bituminous coal mine dusts. J. Aero. Sci. 28, 689–696.

Harrison J., J.-Q. Chen, W. Miller, W. Chen, E. Hnizdo, J. Lu, W. Chisholm, M. Keane, P. Gao & W.E. Wallace, 2005. Risk of silicosis in cohorts of Chinese tin and tungsten miners and pottery workers (II): Workplace-specific silica particle surface composition. Am. J. Ind. Med. 48, 10–15.

Hayakawa K., A. Nakamura, N. Terai, R. Kizu & K. Ando, 1997. Nitroarene concentrations and direct-acting mutagenicity of diesel exhaust particulates fractionated by silica-gel column chromatography. Chem. Pharm. Bull. 45((11), 1820–1822.

Heinrich U., H. Muhle, S. Takenaka, E. Ernst, R. Fuhst, U. Mohr, F. Pott & W. Stober, 1986. Chronic effects on the respiratory tract of hamsters, mice, and rats after long-term inhalation of high concentrations of filtered and unfiltered diesel engine emissions. J. Appl. Toxicol. 6, 383–395.

Heinrich U., D.L. Dungworth, F. Pott, L. Peters, C. Dasenbrock, K. Levsen, W. Koch, O. Creutzenberg & A.T. Schulte, 1994. The carcinogenic effects of carbon black particles and tar-pitch condensation aerosol after inhalation exposure of rats. Ann. Occupat. Hyg. 38(S1), 351–356.

Hill C.A., W.E. Wallace, M.J. Keane & P.S. Mike, 1995. The enzymatic removal of a surfactant coating from quartz and kaolin by P388D1 cells. Cell Biol. Toxicol. 11, 119–128.

Hnizdo V. & W.E. Wallace, 2002. Monte carlo analysis of the detection of clay occlusion of quartz particles using multiple voltage SEM-X-ray spectroscopy. Scanning 24, 264–269.

Hussain S.M., K.L. Hess, J.M. Gearhart, K.T. Geiss & J.J. Schlager, 2005. In vitro toxicity of nanoparticles in BRL 3A rat liver cells. Toxicol. In vitro 19, 975–983.

International Agency for Research on Cancer., 1989. Diesel and Gasoline Engine Exhausts and Some Nitroarenes. Vol. 46. Monographs on the Evaluation of Carcinogenic Risks to Humans. Lyon, France: World Health Organization, IARC.

Jaurand M.C., L. Magne & J. Bignon, 1979. Inhibition by phospholipids of hemolytic action of asbestos. Br. J. Ind. Med. 36, 113–116.

Johnson N.F. & K.R. Maples, 1994. In: Fiber-induced Hydroxyl Radical Formation and DNA Damage; Cellular and Molecular Effects of Mineral Dusts and Fibres NATO ASI Series H, Cell Biology, Vol. 85. Springer-Verlag, Berlin, Heidelberg, pp. 23–37.

Keane M.J., W.E. Wallace, M. Seehra, C. Hill, V. Vallyathan, P. Raghootama & P. Mike, 1990. In: Respirable Particulate Interactions with the Lecithin Component of Pulmonary Surfactant; VII International Pneumoconiosis Conference. DHHS (NIOSH) Publication #90-108 Part 1, pp. 231–244.

Keane M.J., S.G. Xing, J. Harrison, T. Ong & W.E. Wallace, 1991. Genotoxicity of diesel exhaust particles dispersed in simulated pulmonary surfactant. Mutat. Res. 260, 233–238.

Keane M.J., J.L. Hornsby-Myers, J.W. Stephens, J.C. Harrison, J.R. Myers & W.E. Wallace, 2002a. Characterization of hard metal dusts from sintering and detonation coating processes and comparative hydroxyl radical production in vitro. Chem. Res. Toxicol. 15, 1010–1016.

Keane M.J., J. Martin, J. Hornsby-Myers, J. Stephens, J. Harrison, J. Myers, T. Ong & W. Wallace, 2002b. Particle characterization, free radical generation, and genotoxicity of hard metal and detonation coating dusts. In: Proceedings of

Inhaled Particles IX. Cambridge UK. Ann. Occupat. Hyg. 46(S1):402–405.

Keane M. & W. Wallace, 2005. A quantitative in vitro fluorescence imaging method for phospholipid loss from respirable mineral particles. Inhal. Toxcol. 7, 287–292.

Kendall M., L. Brown & K. Trought, 2004. Molecular adsorption at particle surfaces. Inhalation Tox. 16(S1), 99–105.

King L.C., M.J. Kohan, A.C. Austin, L.D. Claxton & J.L. Hunsingh, 1981. Evaluation of the release of mutagens from diesel particles in the presence of physiological fluids. Environ. Mutag. 3, 109–129.

Kozin F. & B.J. McCarty, 1977. Protein binding to monosodium urate monohydrate, calcium pyrophosphate dihydrate, and silicon dioxide crystals. I. Physical characteristics. J. Lab. Clin. Med. 89, 1314–1325.

Kreigseis W. & A. Scharmann, 1982. Specific harmfulness of respirable dusts from West German coal mines V: Influence of mineral surface properties. Ann. Occupat. Hyg. 26, 511–525.

Lam C.W., J.T. James, R. McCluskey & R.L. Hunter, 2004. Pulmonary toxicity of carbon nanotubes in mice 7 days and 90 days after intratracheal instillation. Toxicol. Sci. 77, 126–134.

Lansne C., Z.W. Gu, W. Venegas & I. Chouroulinkov, 1984. The in vitro micronucleus assay for detection of cytogenetic effects induced by mutagen-carcinogens: Comparison with the in vitro sister-chromatid exchange assay. Mutat. Res. 130, 273–282.

LeBouffant L., H. Daniel, J.C. Martin & S. Bruyere, 1982. Effect of impurities and associated minerals on quartz toxicity. Ann. Occupat. Hyg. 26, 625–634.

Liu X., M.J. Keane, B.Z. Zhong, T. Ong & W.E. Wallace, 1996. Micronucleus formation in V79 cells treated with respirable silica dispersed in medium and in simulated pulmonary surfactant. Mutat. Res. 361, 89–94.

Liu X., M.J. Keane, J.C. Harrison, E.V. Cilento, T. Ong & W.E. Wallace, 1998. Phospholipid surfactant adsorption by respirable quartz and in vitro expression of cytotoxicity and DNA damage. Toxicol. Lett. 96(7), 77–84.

Lu W., Y. Zhang, Y.Z. Tan, K.I. Hu, X.G. Jiang & S.K. Fu, 2005. Cationic albumin-conjugated pegylated nanoparticles as novel drug carrier for brain delivery. J. Control Release 107, 428–448.

Mauderly J., R.K. Jones, W.C. Griffith, R.F. Henderson & F.R.O. McClellan, 1987. Diesel exhaust is a pulmonary carcinogen in rats exposed chronically by inhalation. Fundam. Appl. Toxicol. 9, 208–221.

Maynard A.D. & E.D. Kuempel, 2005. Airborne nanoparticles and occupational health. J. Nanoparticle Res. 7, 587–614.

McClellan R.O., A.L. Brooks, R.G. Cuddihy, R.K. Jones, J.L. Mauderly & R.K. Wolff, 1982. Inhalation toxicology of diesel exhaust particles. Dev. Toxicol. Environ. Sci. 10, 99–120.

McMillian M.H., M. Cui, M. Gautam, M. Keane, T. Ong, W. Wallace & E. Robey, 2002. Mutagenic potential of particulate matter from diesel engine operation on Fischer-Tropsch fuel as a function of engine operating conditions and particle size. Soc. Auto. Engineers Technical Paper 2002-01-1699, pp. 1–18.

Murray D., J. Harrison & W. Wallace, 2005. A ^{13}C CP/MAS and ^{31}P NMR study of the interactions of dipalmitoyl phosphatidyl choline with respirable silica and kaolin. J. Colloid Interface Sci. 288, 166–170.

Michalet X., F.F. Pinaud, L.A. Bentolila, J.M. Tsay, S. Doose, J.J. Li, G. Sundaresan & A.M. Wu, 2005. Quantum dots for live cells, in vivo imaging and diagnostics. Science 307, 538–544.

Mitchell A.D., D.A. Casciano, M.L. Meltz, D.E. Robinson, R.H.C. San, G.M. William & E.S. Von Halle, 1983. Unscheduled DNA synthesis tests: A report of the U. S. Environmental Protection Agency Gene-Tox Program. Mutat. Res. 123, 363–410.

Morimoto K., M. Kitamura, H. Kondo & A. Korizumi, 1986. Genotoxicity of diesel exhaust emissions in a battery of in-vitro short-term and in-vivo bioassays. Dev. Toxicol. Envir. Sci. 13, 85–101.

National Institute for Occupational Safety and Health., 1988. Carcinogenic Effects of Exposure to Diesel Exhaust NIOSH Current Intelligence Bulletin 50 DHHS (NIOSH) Publication #88-116. Atlanta, GA: Centers for Disease Control and Prevention.

Nikula K.J., M.B. Snipes, E.B. Barr, W.C. Griffith, R.F. Henderson & J.L. Mauderly, 1995. Comparative pulmonary toxicities and carcinogenicities of chronically inhaled diesel exhaust and carbon black in F344 rats. Fundam. Appl. Toxicol. 25(1), 80–94.

Oberdorster E., 2004. Manufactured nanomaterials (fullerenes, C60) induce oxidative stress in the brain of juvenile largemouth bass. Environ. Health Perspect. 112(10), 1058–1062.

Oberdörster G., R. Ferin, J. Gelein, S.C. Soderholm & J. Finkelstein, 1992. Role of alveolar macrophage in lung injury – studies with uktrafine particles. Envir. Health Persp. 97, 193–199.

Oberdörster G., R.M. Gelein, J. Ferin & B. Weiss, 1995. Association of particulate air pollution and acute mortality: Involvement of ultrafine particles?. Inhal. Toxicol. 7, 111–124.

Oberdörster G., J. Ferin, S. Soderholm, R. Gelein, C. Cox, R. Baggs & P.E. Morrow, 1994. Increased pulmonary toxicity of inhaled due to lung overload alone?. Ann. Occupat. Hyg. 38(S1), 295–302.

Oberdörster G., 2000. Toxicology of ultrafine particles: in vivo studies. Phil. Trans. Roy. Soc. London Ser. A 358(1775), 2719–2740.

Oberdörster G., Z. Sharp, V. Atudorei, A. Elder, R. Gelein, W. Kreyling & C. Cox, 2004. Translocation of inhaled ultrafine particles to the brain. Inhal. Toxicol. 16(6–7), 437–445.

Pantarotto D., C.D. Partidos & J. Hoebeke, 2003. Immunization with peptide-functionalized carbon nanotubes enhances virus-specific neutralizing antibody responses. Chem. Biol. 10(10), 961–966.

Pantarotto D., J.P. Briand & M. Prato, 2004a. Translocation of bioactive peptides across cell membranes by carbon nanotubes. Chem. Commun. 1, 16–17.

Pantorotto D., R. Singh & D. McCarthy, 2004b. Functionalized carbon nanotubes for plasmid DNA gene delivery. Angew. Chem. Int. Edit. 43(39), 5242–5246.

Perry P. & S. Wolff, 1974. New Giemsa method for the differential staining of sister chromatids. Nature (London) 251, 156–158.

Peters K., R.E. Unger, C.J. Kirkpatrick, A.M. Gatti & E. Monari, 2004. Effects of nano-scaled particles on endothelial cell function in vitro: Studies on viability, proliferation and inflammation. J. Mater. Sci. Mater. Med. 15(4), 321–325.

Robock K. & W. Klosterkotter, 1973. Investigations into the specific toxicity of different SiO_2 and silicate dusts. Staubreinhalt. Luft. 33, 60–63.

Sadeghiani N., L.S. Barbosa, M.H.A. Guedes, S.B. Chavez, J.G. Santos, O. Silva, F. Pelegrini, A.B. Azevedo, P.C. Morais & Z.G.M. Lacava, 2005. Magnetic resonance of polyaspartic acid-coated magnetite nanoparticles administered in mice. IEEE Trans. Magn. 41, 4108–4110.

Sayes C.M., A.M. Gobin, K.D. Ausman, J. Mendez, J.L. West & V.L. Colvin, 2005. Nano-C-60 cytotoxicity is due to lipid peroxidation. Biomaterials 26, 7587–7595.

Saxena Q.B., R.K. Saxena, P.D. Siegel & D.M. Lewis, 2003. Identification of organic fractions of diesel exhaust particulate (DEP) which inhibit nitric oxide (NO) production from a murine macrophage cell line. Toxicol. Lett. 143, 317–322.

Scarpelli E.M., 1968 The Surfactant System of the Lung. Philadelphia: Lea and Febiger 97.

Siegel P.D., R.K. Saxena, Q.B. Saxena, J.K.H. Ma, J.Y.C. Ma, X.-J. Yin, V. Castronova, N. Al-Humadi & D.M. Lewis, 2004. Effect of diesel exhaust particulate (DEP) on immune responses: Contributions of particulate versus organic soluble components. J. Toxicol. Environ. Health 67(A), 221–231.

Shvedova A.A., E.R. Kisin, A.R. Murray, D. Schwegler-Berry, V.Z. Gandelsman, A. Maynard, P. Baron & V. Castranova, 2003. Exposure to carbon nanotube material: Assessment of the biological effects of nanotube materials using human keratinocytes. J. Toxcol. Environ. Health 66, 1901–1918.

Shvedova A.A., E.R. Kisin, R. Mercer, A.R. Murray, V.J. Johnson, A.I. Potapovich, Y.Y. Tyurina, O. Gorelik, S. Arepalli, D. Schwegler-Berry, A.F. Ubbs, J. Antonini, D.E. Evans, B-K. Ku, D. Ramsey, A. Maynard, V. KaganV.E. Castranova & P. Brown, 2005. Unusual inflammatory and fibrogenic pulmonary response to single-walled carbon nanotubes in mice. Am. J. Physiol. Lung Cell Mol. Physiol. 289, L698–L708.

Slauson D.O. & B.J. Cooper, 1990. Mechanisms of Disease (2nd ed.). Wilkins Boltimore: Williams 192–193.

Soontjens C.D., K. Holmberg, R.N. Westerholm & J.J. Rafter, 1997. Characterization of polycyclic aromatic compounds in diesel exhaust particulate extract responsible for aryl hydrocarbon receptor activity. Atmosp. Environ. 31(2), 219–225.

Soto K.F., A. Carrasco, T.G. Powell, K.M. Garza & L.E. Murr, 2005. Comparative in vitro cytotoxicity assessment of some manufactured nanoparticulate materials characterized by transmission electron microscopy. J. Nanopart. Res. 7, 145–169.

Stephens J.W., J.C. Harrison & W.E. Wallace, 1998. Correlating Auger electron spectroscopy with scanning electron microscopy – energy dispersive spectroscopy for the analysis of respirable particles. Scanning 20, 302–310.

Tao F., B. Gonzalez-Flecha & L. Kobnik, 2003. Reactive oxygen species in pulmonary inflammation by ambient particulates. Free Rad. Biol. Med. 35(4), 327–340.

Tran C.L., D. Buchanan, R.T. Cullen, A. Searl, A.D. Jones & K. Donaldson, 2000. Inhalation of poorly soluble particles. II. Influence of particle surface area on inflammation and clearance. Inhal. Toxicol. 12, 1113–1126.

Vallyathan V., D. Schwegler, M. Reasor, L. Stettler, J. Clere & F.H.Y. Green, 1988. Comparative in vitro cytotoxicity and relative pathogenicity of mineral dusts. Ann. Occupat. Hyg. 32, 279–289.

Wallace W.E., L.C. Headley & K.C. Weber, 1975. Dipalmitoyl lecithin adsorption by kaolin dust in vitro. J Colloid Interface Sci. 5, 535–537.

Wallace W.E., V. Vallyathan, M.J. Keane & V. Robinson, 1985. In vitro biological toxicity of native and surface modified silica and kaolin. J. Toxicol. Environ. Health 16, 415–424.

Wallace W.E., M.J. Keane, C.A. Hill, J. Xu & T. Ong, 1987. Mutagenicity of diesel exhaust particles and oil shale particles dispersed in lecithin surfactant. J. Toxicol. Environ. Health 21, 163–171.

Wallace W.E., M.J. Keane, V. Vallaythan, P. Hathaway, E.D. Regad, V. Castranova & F.H.Y. Green, 1988. Suppression of inhaled particle cytotoxicity by pulmonary surfactant and retoxification by phospholipase. Ann. Occupat. Hyg. 32(1), 291–298.

Wallace W.E., M.J. Keane, P.S. Mike, C.A. Hill & V. Vallyathan, 1989. In: Mossman B.T. and Begin R.O. eds. Mineral Surface-specific Differences in the Adsorption and Enzymatic Removal of Surfactant and their Correlation with Cytotoxicity; Effects of Mineral Dusts on Cells. NATO ASI Series, Vol. H30, Springer Verlag, pp. 49–56.

Wallace W.E., M. Keane, S. Xing, J. Harrison, M. Gautam & T. Ong, 1990a. Mutagenicity of Diesel Exhaust Soot Dispersed in Phospholipid Surfactants. In: Seemayer N.H., & Hadnagy W. ed. Environmental Hygiene II. Springer Verlag, Berlin, pp. 7–10ISBN 0-387-52735-4.

Wallace W.E., J. Harrison, M.J. Keane, P. Bolsaitis, D. Eppelsheimer, J. Poston & S.J. Page, 1990b. Clay occlusion of respirable quartz particles detected by low voltage scanning electron microscopy – X-Ray analysis. Ann. Occupat. Hyg. 34, 195–204.

Wallace W.E., M.J. Keane, P.S. Mike, C.A. Hill, V. Vallyathan & E.D. Regad, 1992. Contrasting respirable quartz and kaolin retention of lecithin surfactant and expression of membranolytic activity following phospholipase A2 digestion. J. Toxicol. Environ. Health 37, 391–409.

Wallace W.E. & M.J. Keane, 1993. Differential surface composition analysis by multiple-voltage electron beam X-ray microscopy. US Patent 5,210,414 (renewed 2000). US Patent Office, Wash. DC.

Wallace W.E., J. Harrison, R.L. Grayson & M.J. Keane, 1994a. Aluminosilicate surface contamination of respirable quartz

38

particles from coal mine dusts and from clay works dusts. Ann. Occupat. Hyg. 38(1), 439–445.

Wallace W.E., M.J. Keane, J.C. Harrison, J.W. Stephens, P.S. Brower, R.L. Grayson, V. Vallyathan & M.D. Attfield, 1994b. Surface properties of respirable silicate and alumino-silicate dusts affecting bioavailability. In: Davis J.M.G., & Jaurand M.C., eds. Cellular and Molecular Effects of Mineral and Synthetic Dusts and Fibres. NATO ASI Series, Vol. H85. Springer-Verlag, Berlin Heidelberg, pp. 369–379.

Walton W.H., J. Dodgson, G.G. Hadden & M. Jacobsen, 1971. The Effect of Quartz and Other Non-coal Dusts in Coal-workers Pneumoconioses. In: Walton W.H. ed. Inhaled Particles IV. 2 Pergamon Press, Oxford, pp. 669–689.

Warheit D.B., B.R. Laurence, K.L. Reed, D.H. Roach, G.A.M. Reynolds & T.R. Webb, 2004. Comparative pulmonary toxicity assessment of single-wall carbon nanotubes in rats. Toxicol. Sci. 77(1), 117–125.

US Environmental Protection Agency, 2002. Health Assessment Document for Diesel Engine Exhaust. USEPA EPA/600/8-90/057F. 01 May (2002) U.S. EPA, Office of Research and Development, National Center for Environmental Assessment, Washington, DC.

Zhi P.X., H.G. Qing, Q.L. Gao & B.Y. Ai, 2005. Inorganic nanoparticles as carriers for efficient drug delivery. Chem Eng. Sci. 61, 1027–1040.

Journal of Nanoparticle Research (2007) 9:39–52
DOI 10.1007/s11051-006-9174-6

Special issue: Nanoparticles and Occupational Health

Plasma synthesis of semiconductor nanocrystals for nanoelectronics and luminescence applications

Uwe Kortshagen*, Lorenzo Mangolini and Ameya Bapat
*Mechanical Engineering, University of Minnesota, 111 Church St. SE, Minneapolis, MN, 55455, USA;
Author for correspondence (E-mail: uk@me.umn.edu)

Received 21 August 2006; accepted in revised form 24 August 2006

Key words: silicon nanocrystals, nanoelectronics, luminescence, plasma reactor

Abstract

Functional nanocrystals are widely considered as novel building blocks for nanostructured materials and devices. Numerous synthesis approaches have been proposed in the solid, liquid and gas phase. Among the gas phase approaches, low pressure nonthermal plasmas offer some unique and beneficial features. Particles acquire a unipolar charge which reduces or eliminates agglomeration; particles can be electrostatically confined in a reactor based on their charge; strongly exothermic reactions at the particle surface heat particles to temperatures that significantly exceed the gas temperature and facilitate the formation of high quality crystals. This paper discusses two examples for the use of low pressure nonthermal plasmas. The first example is that of a constricted capacitive plasma for the formation of highly monodisperse, cubic-shaped silicon nanocrystals with an average size of 35 nm. The growth process of the particles is discussed. The silicon nanocubes have successfully been used as building blocks for nanoparticle-based transistors. The second example focuses on the synthesis of photoluminescent silicon crystals in the 3–6 nm size range. The synthesis approach described has enabled the synthesis of macroscopic quantities of quantum dots, with mass yields of several mg/hour. Quantum yields for photoluminescence as high as 67% have been achieved.

Introduction

Nanoparticles have attracted significant interest due to many novel, size-tunable properties including their size-dependent band gap and photoluminescence (PL) emission (Brus, 1991; Alivisatos, 1996), reduced melting temperatures (Goldstein et al., 1992; Shi, 1994; Zhang et al., 2001), and increased hardness compared to bulk material (Gerberich et al., 2003). A variety of novel devices based on nanoparticles have been studied including light emitting diodes (Colvin et al., 1994; Dabbousi et al., 1995), photovoltaic cells (O'Regan & Grätzel, 1991), nanoparticle based memory devices (Tiwari et al., 1996a, b; Ostraat et al., 2001a, b), single

electron transistors (Klein et al., 1997; Fu et al., 2000; Takahashi et al., 2000; Kim et al., 2002), and gas sensors (Volkening et al., 1995; Holtz et al., 1996; Kennedy et al., 2003; Kennedy et al., 2003). Among the nanoparticle materials studied nowadays, crystalline silicon nanoparticles are of great interest for electronic applications such as single electron transistors (Fu et al., 2000), vertical transistors (Nishiguchi & Oda, 2000), and floating gate memory devices (Tiwari et al., 1996a, b; Ostraat et al., 2001b; Banerjee et al., 2002). Intense research is also performed in the area of silicon nanocrystal-based photonic devices (Littau et al., 1993; Collins et al., 1997; Nayfeh et al., 1999; St. John et al., 1999; Canham, 2000; Ledoux et al., 2000; Borsella et al., 2001; Holmes et al., 2001; Park et al., 2001; Franzò et al., 2002; Ledoux et al., 2002; Pettigrew et al., 2003). Contrary to bulk silicon, strong PL has been observed from silicon nanocrystals even at room temperature (Canham, 2000), since the band gap of silicon nanocrystals becomes more direct and widens significantly at particle sizes of less than 5 nm (Brus et al., 1995; Puzder et al., 2002; Puzder et al., 2003; Zhou et al., 2003a, b; Draeger et al., 2004). Additional advantages of silicon nanocrystals include the element's low toxicity (at least in its bulk form) as compared to many of the compound semiconductors and the existence of a large silicon technology infrastructure.

A wide spectrum of synthesis methods for silicon nanocrystals is already known. In the liquid phase, small silicon crystals have been prepared from porous silicon (Canham, 1990), which is produced by anodizing silicon wafers in hydrofluoric acid solution (Canham, 1990, 2000; Nayfeh et al., 1999; Nayfeh et al., 2001). Other liquid phase processes include the synthesis in inverse micelles (Wilcoxon & Samara, 1999; Wilcoxon et al., 1999), the synthesis in high temperature supercritical solutions (Holmes et al., 2001; Ding et al., 2002), the oxidation of metal silicide (Pettigrew et al., 2003), and the reduction of silicon tetrahalides and other alkylsilicon halides (Baldwin et al., 2002). While solution based processes offer a number of advantages that include their ability to produce particles with a rather narrow size distribution and their capability to cap the nanocrystal surfaces with organic molecules in order to protect them from oxidation, they are often afflicted with rather low production rates.

Aerosol processes for the synthesis of nanoparticles are attractive due to the high processing rates that can be achieved through direct gas to particle conversion. A popular method for the formation of silicon nanocrystals is the high temperature thermal reaction (pyrolysis) of silane in furnace flow reactors (Littau et al., 1993; Onischuk et al., 2000; Ostraat et al., 2001a). This method is capable of high rate throughput, however, like most aerosol processes it suffers from problems of particle agglomeration (Onischuk et al., 2000) which is difficult to avoid when the majority of particles are neutral. Other gas phase methods capable of high throughput include the decomposition of silane or disilane through laser light irradiation (photolysis) (Batson & Heath, 1993) and laser

pyrolysis using high power infrared lasers (Ehbrecht & Huisken, 1999; Ledoux et al., 2000, 2002; Huisken et al., 2003; Li et al., 2003), which both lead to rapid particle formation. Swihart and coworkers recently reported a high rate laser pyrolysis process which produced luminescent silicon nanoparticles at a rate of up to 200 mg/h (Li et al., 2003). However, like with most aerosol approaches, particle agglomeration is a problem. For instance, Borsella et al. (Borsella et al., 2001) report a particle size distribution with particle sizes ranging between 1 and 100 nm for their laser pyrolysis process. In addition, not all the silane is necessarily converted to crystalline particles, as the group of Swihart reports a mixture of amorphous and crystalline material in their TEM studies (Li et al., 2003).

Nonthermal plasmas are characterized by a strong non-equilibrium between the background gas temperature, which remains close to room temperature, and the electron temperature, which is typically around 2 eV (1 eV≈11,000 K). In thermal plasmas, on the other hand, heavy particles and electrons have the same temperature. Nonthermal plasmas offer a number of unique advantages for the synthesis of nanocrystals, which have so far remained largely unrecognized by the aerosol community. Among these desirable attributes are:

(1) Particles immersed in plasmas are usually unipolarly negatively charged (Goree, 1994; Matsoukas & Russel, 1995; Schweigert & Schweigert, 1996; Kortshagen & Bhandarkar, 1999), based on the much higher mobility of electrons in the plasma as compared to that of ions. This *unipolar charge prevents or strongly reduces particle agglomeration*) (Schweigert & Schweigert, 1996; Matsoukas, 1997; Kortshagen & Bhandarkar, 1999).

(2) Negatively charged *nanoparticles can be confined in the plasma reactor*, whose walls are also negatively charged as a result of ambipolar diffusion of electrons and ions (Selwyn et al., 1990; Carlile et al., 1991; Bouchoule & Boufendi, 1993; Selwyn et al., 1993; Boufendi & Bouchoule, 1994; Shen et al., 2003).

(3) Particles immersed in a low pressure plasma experience strong heating through electron-ion and chemical recombination at the particle surface. Combined with the ineffective cooling

of particles at low pressures, this can lead to *particle temperatures that exceed the temperature of the surrounding gas by several hundreds of Kelvin* (Bapat et al., 2004; Mangolini et al., 2005). This process is believed to be responsible for the formation of crystalline nanoparticles at low gas temperatures and the formation of highly oriented, facetted particles.

In this paper, we discuss two examples of low pressure plasma processes used to synthesize silicon nanocrytals. The first process focuses on the synthesis of particles with well-defined morphology several tens of nanometers in size. These particles are used as building blocks for the fabrication of nanoelectronic devices. The second example focuses on the synthesis of silicon nanocrystals in the 3–6 nm range for their luminescent properties. We discuss that plasma synthesis enables high mass yields and the formation of quantum dots with unprecedented quantum yield for silicon.

Silicon nanocrystals for nanoelectronic devices

Experimental setup
The work reported here is based on a capacitively coupled plasma, which is deliberately operated in a mode that causes the plasma to constrict into a rotating filament. The experimental setup for the process is as shown in Figure 1. The plasma is produced in a reactor that consists of a 5 cm inner diameter, ~60 cm long glass chamber. Nanoparticles that are formed in this reactor are extracted by the gas flow and injected into a high vacuum chamber through a 1-mm-orifice. The plasma is

produced by applying ~80–200 W of radiofrequency (RF) power at 13.56 MHz to a ring electrode placed about 15 cm upstream of the orifice. The orifice plate serves as the ground electrode. The discharge is operated in 5% silane diluted in helium and argon at a pressure of 200–270 Pa. Flow rates are typically ~6 sccm, leading to residence times in the reactor of several seconds for the particles and the gas. The pressure in the chamber downstream of the orifice is ~0.13 Pa during plasma operation. Due to the high pressure difference, the gas flow through the orifice is choked. A supersonic gas jet is formed, that expands into the high vacuum chamber. Particles extracted from the plasma are accelerated in the jet to velocities of up to 250 m/s. Particles are then deposited on various substrates either for TEM analysis or for device fabrication.

Particle size and morphology
The particles extracted from the plasma are studied by transmission electron microscopy (TEM) using JEOL 1210 microscope operating at 120 kV accelerating voltage (1 k × 1 k CCD camera). Images of a large number of particles and a higher resolution image of a single particle are shown in Figure 2. The silicon particles are single crystals with a predominantly cubic shape. Electron diffraction confirms that the structure is the normal diamond cubic silicon structure and that the particle faces uniformly are (100) crystal facets of silicon. The particle size distribution is relatively monodisperse and best fit by a Gaussian distribution with an average size of 35 nm and a standard deviation of 4.7 nm. The higher magnification image of a single particle in Figure 2b

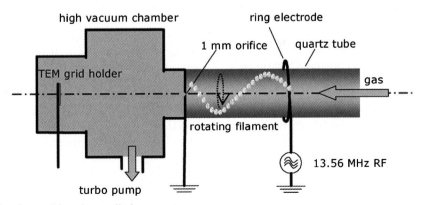

Figure 1. Constricted capacitive plasma discharge system.

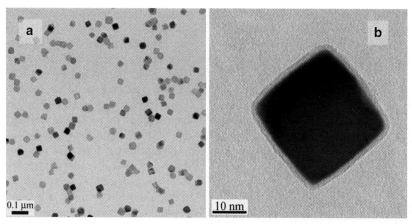

Figure 2. Single crystal, cubic-shaped silicon nanoparticles produced in a nonthermal, constricted capacitive plasma. Experimental setup and conditions are discussed in the text. (a) Low magnification overview, (b) high magnification image of a single particle.

shows an amorphous shell on the particle surface of about 2 nm thickness. This amorphous layer is most likely a native silicon oxide, which grows quickly on exposure of the particles to air. Fourier transform infrared (FTIR) spectroscopy of particles taken just minutes after the production of the particles suggests that the particle surface is initially hydrogen terminated. Silicon oxygen peaks in the initial FTIR spectrum are weak. After exposure of the particles to air for a few hours up to one day the silicon oxygen peaks grow strongly suggesting the formation of a silicon oxide shell.

The cubic shape of silicon nanocrystals is highly unusual; we are not aware of other reports of cubic shaped silicon nanocrystals in the literature. The equilibrium shape of silicon is usually believed to be a shape that features large (111) facets (Eaglesham et al., 1993), since the (111) surface is the lowest energy surface for silicon. However, recent theoretical studies indicate that for hydrogen terminated silicon surfaces the (100) surface may be the minimum energy surface (Stekolnikov et al., 2002). As pointed in (Barnard & Zapol, 2004), under this premise the cube is the equilibrium shape for a silicon crystal with a hydrogen terminated surface. Based on these theoretical studies and our observation of strong silicon hydrogen bond peaks in the FTIR spectra of the particles immediately after production, we believe that the plasma creates a hydrogen-rich environment in which particles crystallize and reach their

equilibrium shape under high hydrogen coverage of the particle surface.

Characterization of the synthesis process

A more detailed study of the plasma provides valuable insight into the particle formation process. As already discussed in (Bapat et al., 2004), the plasma consists of two distinct regions, shown in Figure 3. A more intense, non-stationary plasma is observed downstream of the ring electrode. A high speed camera was used to study the structure and dynamics of this plasma filament; a single frame is shown at the bottom of Figure 3. The filament is striated and consists of approximately 10–15 intense globules. High speed movies taken side-on and end-on showed that the filament rotates close to the tube wall with a frequency of approximately 150 Hz. Laser light scattering experiments (Bapat et al., 2004) seemed to indicate that the diffuse upstream plasma is essential in the initial formation of particles, since a region of intense scattering signal was observed ∼2 cm upstream of the ring electrode. This intense scattering was interpreted as being caused by particles that formed in the diffuse plasma and were trapped in electrostatic potential traps formed by the RF sheath close to the powered ring electrode. Laser scattering also indicated that particles escape from the potential traps in the region close to the reactor wall and enter the region of the rotating filament downstream of the ring electrode.

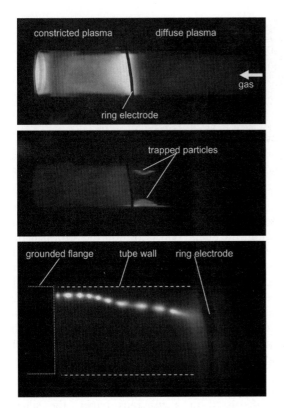

Figure 3. Top: time-averaged image of the discharge showing a dim, diffuse plasma upstream of the ring electrode, and a more intense plasma downstream of the electrode. (Reactor tube pictured here is shorter than the one used in the experiments described in the text). Middle: light scattering signal from a radial scan of a HeNe laser. The signal in the constricted plasma region is due to plasma background radiation and not caused by particles. Bottom: high-speed image of constricted region of the capacitive discharge.

To gain more information about the particle growth in this plasma, particles were extracted along the length of the discharge and the gas composition was analyzed with a quadrupole mass spectrometer. Particles were extracted by inserting a small pyrex probe into the reactor at different axial positions. The extracted gas flow was passed over a TEM grid and particles are collected by diffusion. The Stokes number of a particle is the ratio of the stopping distance to the characteristic dimension of an obstacle and is given by $Stk = \tau U_0/d_{obstacle}$ (Friedlander, 2000), where τ is the relaxation time of the particle, and U_0 the flow velocity. Considering the extraction tube diameter as the characteristic dimension of the obstacle

$d_{obstacle}$, the Stokes number of the particles $Stk \ll 1$ and there are no impaction losses of the particles in the probe. Particles were collected by impaction downstream of the orifice to verify that the usual cubic, monodisperse particles were obtained. We have not found any evidence that the collection mechanism, either by impaction downstream of the plasma or by diffusion after the extraction probe, does have any effect on the particle morphology and shape. At each axial location the gas was also sampled at three radial locations: the tube center, the middle and close to the tube wall. In accordance with the laser scattering results, significantly more particles were observed close to the reactor walls than in the discharge center.

Figure 4 shows typical TEM images of particles collected with the probe close to the reactor wall. At a sampling point 5 cm upstream of the ring electrode (20 cm upstream of the orifice), large amorphous spherical particles with a rough surface are collected, as shown in Figure 4c. The diameter of the particles is between 200 and 400 nm and their structure suggests that they are agglomerates of smaller primary particles. Particles of this kind have been observed in numerous experiments studying particle formation in capacitively coupled silane plasmas (Bouchoule & Boufendi, 1993; Watanabe & Shiratani, 1993; Boufendi & Bouchoule, 1994; Stoffels et al., 1996). Particles collected 2.5 cm upstream of the ring electrode (12.5 cm upstream of the orifice) are shown in Figure 4B. This position corresponds to the strong scattering signal seen in Figure 3 which likely indicates a particle trap. The particles have a similar structure as those further upstream, however, their average size is smaller, ~150-300 nm. Darkfield TEM studies also indicate that particles show signs of poly-crystallinity. Apparently, the exposure to the plasma is favorable for a restructuring of the particles. Whether the particle trapping plays a role in the restructuring is currently unclear. Figure 4a finally shows particles extracted 5 cm downstream of the ring electrode (10 cm upstream of the orifice). The morphology is distinctly different from the particles observed upstream. Particles are significantly smaller, ~40–70 nm diameter, and mostly spherical, however, some first cubic-shaped particles are already observed. The particles are mainly single crystals. A fraction of the particles also shows

44

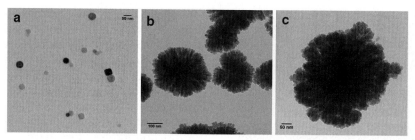

Figure 4. (a) (*5 cm downstream of ring electrode*) 40–70 nm diameter spherical and cubic nanoparticles. (b) (*2.5 cm upstream of ring electrode*) Cauliflower shaped polycrystalline particles. c) (5 cm *upstream of ring electrode*) 200–400 nm diameter cauliflower shaped amorphous particles.

defects such as twin boundaries. Unfortunately, we do currently have no information about the transformation of the particles over the last 10 cm before reaching the orifice. Work to analyze this region is in progress. It should be mentioned that the particles collected downstream of the orifice for all these runs displayed the usual mainly cubic shape.

While we do not currently have an explanation for the significant shrinkage of the particles from the upstream to downstream sampling position, the observation of a significant structural transformations seems to confirm our interpretation given in (Bapat et al., 2004) that the particle growth already occurs in the upstream region, while the downstream region is responsible for particles annealing. As discussed in (Bapat et al., 2004), we estimate the plasma density in the filament globules to be on the order of 10^{12} cm^{-3}. At such high plasma densities, electron-ion recombination at the particle surface should be sufficient to heat the particles to temperatures that exceed the gas temperature by several hundreds of Kelvin. These high temperatures are expected to facilitate the crystalline transformation of the particles. Surface diffusion may also be strongly enhanced allowing the particles to approach their cubic equilibrium shape.

Measurements of the gas composition along the reactor support the hypothesis that the plasma is divided into a growth and an annealing zone. As shown in Figure 5, the density of silane, as indicated by the mass signal of its fragments produced in the mass spectrometer, declines steadily throughout the upstream plasma region. At 20 cm upstream of the orifice (5 cm upstream of the ring electrode), the silane concentration has decreased to only about 5% of its value at the inlet. This is consistent with the observed strong particle

growth in the upstream region. In the downstream region, the silane concentration has dropped further to ~2% of the inlet concentration. It should be noted that the *Peclet number*, the ratio of convective to diffusive flux, for the silane molecules in the reactor is on the order of unity. The drop in the upstream region can thus partly be related to diffusion of silane towards a strong sink, possibly the region closely upstream of the ring electrode where rapid particle growth may occur.

Applications of silicon nanocubes

As discussed in a previous paper, the silicon nanocubes reported here have low electronic defect densities (Campbell et al., 2004). The cubic shape is also favorable for the manufacture of nanoscale devices. When the silicon cubes are deposited on a substrate, they all come down on a (100) surface. The deposited cubes thus all have the same crystallographic orientation. The large flat facets

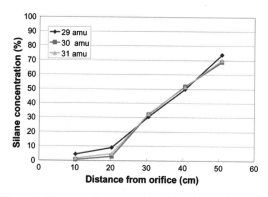

Figure 5. Concentration profiles of various silane fragments obtained by quadrupole mass spectrometry as a function of the axial reactor position. The orifice is at 0 cm, the ring electrode at 15 cm. The 100%-level refers to the signal observed without striking a plasma.

45

are very favorable for electrical contacts. Nanoparticle-based vertical Schottky barrier field effect transistors have been demonstrated based on these particles. This work is reported in a separate publication (Ding et al., 2005).

Luminescent silicon nanoparticles

Silicon in its bulk form is a material with rather poor optical emission and absorption properties due to its indirect band gap, which requires that photon emission and absorption involve a momentum balancing phonon. This fact has so far prevented the development of silicon-based optoelectronic devices, which would have the potential to enable a new level of integration of silicon electronics with optical devices on a single chip. Hence first reports of room temperature light-emission from quantum confined silicon structures were met with great enthusiasm. Initial processes for the production of luminescent silicon nanostructures used magnetron sputtering of silicon in a hydrogen atmosphere (Furukawa & Miyasato, 1988) and the production of porous silicon (Canham, 1990; Cullis & Canham, 1991). A wide range of synthesis approaches has since been proposed both in the liquid (Wilcoxon & Samara, 1999; Holmes et al., 2001; Baldwin et al., 2002; Pettigrew et al., 2003) and in the gas phase (Batson & Heath, 1993; Littau et al., 1993; Ehbrecht & Huisken, 1999; Ostraat et al., 2001a; Li et al., 2003).

However, the synthesis of macroscopic amounts silicon nanocrystals with high optical efficiency has remained a challenge.

Experimental setup

Here we report an approach based on a variation of the reactor discussed above for the synthesis of silicon nanocubes. The main difference from the above approach is that the residence time of particles in the plasma is of the order of a few ms in contrast to the few seconds in the approach above. A schematic of the plasma reactor is shown in Figure 6. An argon-silane gas mixture is passed through a reactor which consists of a 9 mm outer diameter quartz tube with an inner diameter of 6 mm. Two copper rings with an intermediate gap of 6 mm serve as electrodes. The plasma exhibits an intense emission between the electrodes and downstream of the electrode pair. Weaker emission is also observed upstream of the ring electrodes. The residence time of the gas is estimated based on the distance between the grounded ring electrode and the fitting of the quartz tube to the vacuum chamber. This distance is kept constant at 2.5 cm. The plasma is typically generated at a pressure of 186 Pa. The total gas flow rate is adjusted between 12 and 60 sccm, changing the residence between ~7 and ~1.2 ms. The discharge is operated in a SiH_4–Ar mixture and the silane partial pressure is adjusted between 1.3 and 8 Pa. The RF power is delivered to the discharge

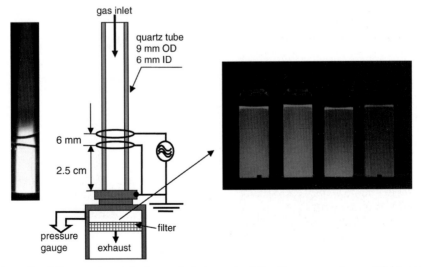

Figure 6. Left: schematic of the experimental setup and photograph of the silane-argon plasma. Right: particles collected on the filter and dispersed in methanol. The excitation wavelength used was 360 nm.

through a T-type matching network, and the discharge is excited at a frequency of 13.56 or 27.12 MHz. Measurements of the RF current and voltage indicate that the power consumption in the plasma is of the order of only a few Watt. Given the very small discharge volume ($< 1 \text{ cm}^3$), a considerable power density is achieved in the plasma region. The particles produced in the plasma are collected on a fine stainless steel mesh placed about 2.5 cm downstream of the discharge. After a few minutes of deposition, the mesh is completely coated with particles, forming a very loosely aggregated deposit. The particles are collected by diffusion, and we have no evidence that the collection mechanism has any influence on the particle morphology.

After deposition, the mesh can be extracted from the system. The silicon nanocrystals exhibit PL as produced, but their luminescence increases on oxidation. Samples of silicon particles that were rinsed into vials with methanol are shown in Figure 6. The particles show strong luminescence in the red when illuminated with a UV light source at 360 nm.

Particle size distribution

For silicon nanocrystals to show strong PL, their size needs to be smaller than ~6 nm (Reboredo et al., 1999). In previous studies, we had found that the PL spectrum correlates well with the residence time of the nanocrystals in the plasma region. The residence time for typical flow rates is between 2 and 6 ms. We found that for shorter residence times the PL emission shifts towards the blue end of the spectrum. According to quantum confinement, this corresponds to smaller nanocrystal sizes (Reboredo et al., 1999). The particle size distribution and the particle crystallinity were studied by TEM analysis for which particles were collected on TEM grids that were placed on the

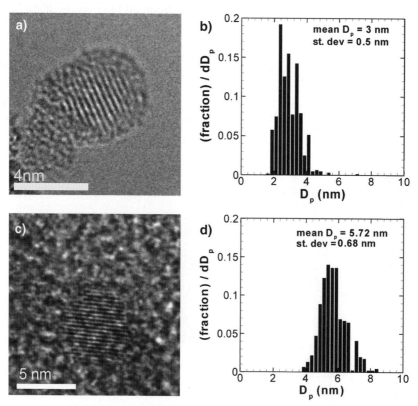

Figure 7. TEM images of the produced particles with particle size distribution as obtained from micrograph analysis. (a) and (b) were obtained for 40 sccm Ar, 8 sccm Ar:SiH$_4$ (95:5); (c) and (d) for 10 sccm Ar, 2 sccm Ar:SiH$_4$ (95:5). (Image (a) courtesy of Dr. C. Perrey taken on a 300 kV Tecnai F30).

particle collection filter. Figure 7 shows high magnification TEM images as well as the particle size distributions for both a short and a long residence time case. For Figure 7a, b the Ar flow rate is 40 sccm and the Ar–SiH_4 (95:5) flow rate is 8 sccm, corresponding to a residence time of 1.85 ms. For Figure 7c, d the Ar flow rate is 10 sccm and the Ar–SiH_4 (95:5) flow rate is 2 sccm; the residence time is 6.8 ms. The partial pressure of silane is the same for these two cases and equals 1.5 Pa.

It can be seen that reducing the residence time leads to a decrease of the particle size from an average of 5.7–3 nm. The size distribution is relatively monodisperse, with a standard deviation of 0.5 nm for the short residence time case, and 0.68 nm for the long residence time case. Particles observed for these two cases are almost exclusively crystalline. Figure 7c shows fringes from the (111) plane of a 5-nm silicon particle deposited on a regular carbon film grid. The particle in Figure 7a has a crystalline core of around 2 nm in diameter and an amorphous layer about 1 nm in thickness which is likely a native oxide. A crystalline fraction of ~100% is observed for the long residence time case. Decreasing the residence time of the particles in the plasma leads to the production of smaller particles, however, the crystalline fraction is often smaller than 100%.

As discussed above, we believe that the particle heating through exothermic surface reactions is the main cause leading to the formation of crystalline particles. In Mangolini et al. (2005) we analyzed the particle heating for the experiments performed here. In this paper, we had measured the plasma density to be $\sim 5 \times 10^{10}$ cm^{-3}. A time-dependent solution of the energy balance for the particles showed that electron-ion recombination at the particle surface could heat a 3-nm particle to temperatures as high as 900 K, even if the surrounding gas was assumed to be at room temperature.

Process mass yield
Achieving macroscopic quantities of silicon nanocrystals has been a challenge for most synthesis approaches. Here we used a combination of mass spectroscopy and direct mass measurements to determine the process yield. Measurements with a mass spectrometer showed that the plasma dissociates virtually 100% of the fed silane precursor. A flow rate of 40 sccm of Argon and 8 sccm of

Ar–SiH_4 (95:5) corresponds to a mass flow rate of silicon of 30 mg/h. The deposition rate on the filter was measured to be 14.4 mg/h. At the same time, a silicon film was deposited in the reactor tube at a rate of 14.7 mg/h. This result indicates that slightly less than 50% of the silane is converted into silicon nanocrystals, while another 50% are deposited as a silicon film on the reactor walls. For slightly larger nanocrystals with emission wavelength peaking at 800 nm we achieved a net yield of ~40 mg/h. To our knowledge, these mass yields exceed other reports in the literature. It should also be pointed out that the process described here is highly scalable and that significantly higher mass yields can likely be achieved by either using a larger reactor or parallelizing smaller reactors.

Influence of surface oxidation
A significant advantage of the plasma approach is that it enables the synthesis of silicon particles in an essentially oxygen-free environment. Silicon is notorious for its tendency to react with oxygen. Unfortunately, the oxidation of silicon nanocrystals is known to severely affect the optical properties by reducing the quantum yield and limiting the optical band gap (Wolkin et al., 1999; Puzder et al., 2002; Vasiliev et al., 2002, 2003a, b).

We studied the effect of oxygen by collecting particles in the vacuum system and performing *in situ* PL studies. Results are shown in Figure 8. The emission spectrum of the nanocrystals still in the vacuum system peaks at 830 nm. On exposure to air, the PL initially shows a strong increase by about a factor of 2.5 without any shift in wave-

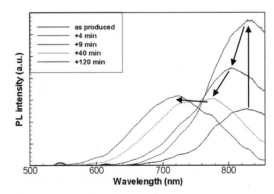

Figure 8. Temporal evolution of the PL spectrum caused by the oxygen exposure of silicon nanocrystals. The initial "as produced" curve corresponds to the emission of the crystals still under vacuum.

length. We interpret this increase as being caused by oxygen reacting at the surface and passivating surface states such as dangling bonds, which are widely believed to reduce the PL efficiency of silicon quantum dots. After just a few minutes of air exposure the PL decreases again and a blue shift of the PL spectrum is observed. This blue shift is likely caused by the formation of a silicon oxide layer at the particle surface, which leads to a shrinking of the silicon crystalline core and to a corresponding widening of the band gap, which is determined by quantum confinement. The decrease in PL emission is consistent with reports in the literature that the quantum yield of oxidized silicon nanocrystals (Littau et al., 1993; Wilcoxon et al., 1999; Li et al., 2003) is usually significantly lower than that of nanocrystals whose surface has been protected from oxidation by functionalization with organic molecules (Holmes et al., 2001; Sankaran et al., 2005).

The hypothesis of surface oxidation discussed above is also supported by FTIR measurements. Samples of as produced particles were studied with FTIR with strong efforts being made to limit the exposure of the particles to air. The FTIR spectrum of as produced particles indicates the presence of silicon–hydrogen bonds and silicon–silicon bonds while a silicon–oxygen signature was virtually absent. Leaving these samples exposed to the atmosphere led to the appearance of strong silicon–oxygen bond related peaks in the FTIR spectrum within a matter of a few hours.

High quantum yield silicon nanocrystals
The fact that the plasma synthesis of silicon nanocrystals is conducted in a virtually oxygen-free environment is very favorable for achieving high quantum yields, if the silicon nanocrystal surface can be passivated without surface oxidation. In a separate publication, we report the surface passivation of silicon nanocrystal with alkene organic molecules under oxygen-free conditions (Jurbergs et al., submitted). Here, we just give a brief summary of the procedure and results. Particles were collected and removed from the system under a dry nitrogen atmosphere in a glove bag and transferred to the nitrogen atmosphere of a glove box for further processing. In a wet-chemical reaction known as hydrosilylation (Buriak, 2002) octadecene molecules were grafted to the nanocrytal surface. After the hydrosilylation

reaction, particles can be suspended in nonpolar solvents and form a clear colloidal solution.

Quantum yields of the passivated silicon quantum dots were measured using an absolute measurement. An integrating sphere was used to simultaneously measure the quantum dot absorption of a blue pump light emitting diode at 380 nm, and the PL in the red range of the spectrum. The emission and absorption were measured with a spectrometer whose spectral response had been calibrated with a NIST traceable calibration lamp. The quantum yield, defined as the number of visible photons emitted per absorbed UV photon, was obtained by integrating over the number of photons in the absorption and emission peaks. The procedure was tested by measuring the quantum yield of dyes and phosphors with known quantum yields and good agreement with these standards was obtained.

Quantum yield measurements of the colloidal silicon quantum dots were performed immediately after the hydrosilylation reaction. Measured quantum yields of the samples are shown in Table 1. Very high quantum yields were achieved for the long wavelength samples. A quantum yield of $67 \pm 7\%$ was measured for the sample with a peak wavelength of 789 nm; $53 \pm 7\%$ were found for the sample peaking at 765 nm. To our knowledge, these are the highest quantum yields for ensembles of silicon particles that have been reported in the literature to date. The quantum yield for samples with smaller peak wavelengths is significantly smaller. This drop in quantum yield is so far not understood. The quantum yields of the same samples were measured again after 8 days, during which the samples were kept in vials that were sealed with caps, but not protected from oxidation by any other means. The measurements of the samples after 8 days suggested that in spite of the surface termination with octadecene some oxidation occurred, since a blue shift of the PL and a significant drop in quantum yield was observed. For comparison, also the quantum yields of several oxidized silicon particle samples were measured. The quantum yields were generally low between 5% and 8%.

In the case of CdSe quantum dots, liquid phase techniques are available for the coating of the particles with an inorganic shell, typically ZnS. This approach is very effective at preventing

Table 1. Experimental conditions and quantum yields (QY) for different silicon quantum dot samples

Ar (sccm)	Ar:SiH$_4$) (95:5) (sccm)	H$_2$ (up) (sccm)	H$_2$ (down) (sccm)	Peak PL wavelength (nm)	QY (%)
60	15	1	50	789	62 ± 11
75	15	1.7	50	765	46 ± 11
85	15	2.6	50	736	22 ± 3
95	15	3.5	50	709	13 ± 2
110	15	4	50	693	1.8 ± 1

H$_2$ (up) refers to the flow of hydrogen injected with the other precursor gas upstream of the plasma. H$_2$ (down) denotes additional hydrogen injected after the plasma. The peak emission wavelength was measured *in situ* in the system before air exposure or any other further processing. The RF power for all experiments was 70 W.

oxidation. No such technique has yet been developed in the case of silicon, either in the liquid phase or in the gas phase, and grafting organic molecules to the particle surface is the best approach available to slow down the growth of a native oxide shell.

Conclusions

In this paper we have discussed nonthermal plasmas as well-defined sources for monodisperse, nonagglomerated nanocrystals. While we have focused here on silicon, we expect that the results presented here can be transferred to a number of other materials systems as well. Two examples for silicon nanocrystal synthesis were presented. The characteristics of the two processes are summarized in Table 2.

In process 1, based on a constricted mode capacitive plasma, cubic-shaped silicon nanocrystals with an average size of 35 nm and a narrow size distribution were synthesized. The nanoparticles were single crystals with well defined (100) facets. This highly unusual shape for silicon crystals was interpreted in terms of the hydrogen termination of the silicon surface, which, according to reports of theoretical studies, makes the

(100)-plane the energetically most favorable surface. The growth process of particles was shown to occur in two distinct phases. In the first stage, particles grow to 200–400 nm size cauliflower-like amorphous particles in the dim diffuse plasma upstream of the ring RF electrode. In the second stage, particles enter the region of the rotating plasma filament, where the particles get transformed into smaller single crystal particles. We identified the heating of the particles by exothermic reactions at the particle surface as a possible mechanism that enables the particles to achieve their equilibrium shape.

With process 2, we presented a process for the synthesis of luminescent silicon nanocrystals. Particles between 3 and 6 nm in size were formed in a simple, single-step flow-through reactor on the time-scale of a few milliseconds. Production yields of luminescent nanocrystals on the order of several tens of milligrams per hour were achieved. The synthesis of nanoparticles in an oxygen-free plasma environment, combined with oxygen-free surface passivation with organic molecules, enabled very high quantum yields of up to 67%. To our knowledge, these quantum yields for ensembles of silicon particles exceed those in other reports in the literature. While the significant drop of the quantum yield toward shorter wavelengths still requires

Table 2. Table summarizing the major differences in the two techniques of particle synthesis

Process	$\tau_{residence}$ (s)	d_p (nm)	Particle shape and morphology	Plasma description	Applications
1	4	35	Single crystal cubes	Constricted, rotating filamentary discharge	Single particle electronic devices
2	$<10^{-2}$	< 5	Crystalline spheres	Uniform discharge	Optoelectronics, photovoltaics

explanation, the plasma approach appears very promising for the synthesis of quantum confined silicon and possibly other quantum dot materials.

Acknowledgments

This work was supported in part by the National Science Foundation under MRSEC award number DMR-0212302, under NIRT-grant DMI-0304211, grant CTS-0500332 and under IGERT award number DGE-0114372, and by InnovaLight, Inc. We acknowledge Dr. Christopher R. Perrey and Professor C. Barry Carter for support with high-resolution TEM.

References

Alivisatos A.P., 1996. Semiconductor clusters, nanocrystals, and quantum dots. Science 271(5251), 933–937.

Baldwin R.K., K.A. Pettigrew, J.C. Garno, P.P. Power, G.-Y. Liu & S.M. Kauzlarich, 2002. Room temperature solution synthesis of alkyl-capped tetrahedral shaped silicon nanocrystals. J. Am. Chem. Soc. 124(7), 1150–1151.

Banerjee S., S. Huang, T. Yamanaka & S. Oda, 2002. Evidence of storing and erasing of electrons in a nanocrystalline-Si based memory device at 77 K. J. Vac. Sci. Technol. B 20(3), 1135–1138.

Bapat A., C. Anderson, C.R. Perrey, C.B. Carter, S.A. Campbell & U. Kortshagen, 2004. Plasma synthesis of single-crystal silicon nanoparticles for novel electronic device applications. Plasma Phys. Controlled Fusion 46(12), B97–B109.

Barnard A. & P. Zapol, 2004. A model for the phase stability of arbitrary nanoparticles as a function of size and shape. J. Chem. Phys. 121(9), 4276–4283.

Batson P.E. & J.R. Heath, 1993. Electron energy loss spectroscopy of single silicon nanocrystals: the conduction band. Phys. Rev. Lett. 71(6), 911–914.

Borsella E., M. Falconieri, S. Botti, S. Martelli, F. Bignoli, L. Costa, S. Grandi, L. Sangaletti, B. Allieri & L. Depero, 2001. Optical and morphological characterization of Si nanocrystals/silica composites prepared by sol–gel processing. Mater. Sci. Eng. B: Solid-State Mater. Adv. Technol. B 79(1), 55–62.

Bouchoule A. & L. Boufendi, 1993. Particulate formation and dusty plasma behaviour in argon-silane RF discharge. Plasma Sources Sci. Technol. 2, 204.

Boufendi L. & A. Bouchoule, 1994. Particle nucleation and growth in a low-pressure argon-silane discharge. Plasma Sources Sci. Technol. 3, 263.

Brus L.E., 1991. Quantum crystallites and nonlinear optics. Appl. Phys. A 53, 465–474.

Brus L.E., P.J. Szajowski, W.L. Wilson, T.D. Harris, S. Schuppler & P.H. Citrin, 1995. Electronic spectroscopy and photophysics of Si nanocrystals: relationship to bulk c-Si and porous Si. J. Am. Chem. Soc. 117, 2915–2922.

Buriak J.M., 2002. Organometallic chemistry on silicon and germanium surfaces. Chem. Rev. 102(5), 1271–1308.

Campbell, S.A., U. Kortshagen, A. Bapat, Y. Dong, S. Hilchie & Z. Shen, 2004. The Production and electrical characterization of free standing cubic single crystal silicon nanoparticles. J. Mater 56(10), 26–28.

Canham L., 2000. Gaining light from silicon. Nature 408, 411–412.

Canham L.T., 1990. Silicon quantum wire array fabrication by electrochemical and chemical dissolution of wafers. Appl. Phys. Lett. 57, 1046.

Carlile R.N., S. Geha, J.F. O'Hanlon & J.C. Stewart, 1991. Electrostatic trapping of contamination particles in a process plasma environment. Appl. Phys. Lett. 59, 1167.

Collins R.T., P.M. Fauchet & M.A. Tischler, 1997. Porous silicon: from luminescence to LEDs. Phys. Today 50, 24.

Colvin V.L., M.C. Schlamp & A.P. Alivisatos, 1994. Light-emitting-diodes made from cadmium selenide nanocrystals and a semiconducting polymer. Nature 370(6488), 354–357.

Cullis A.G. & L.T. Canham, 1991. Visible light emission due to quantum size effects in highly porous crystalline silicon. Nature 335, 335–338.

Dabbousi B.O., M.G. Bawendi, O. Onitsuka & M.F. Rubner, 1995. Electroluminescence from Cdse quantum-dot polymer composites. Appl. Phys. Lett. 66(11), 1316–1318.

Ding, Y., Y. Dong, A. Bapat, J. Deneen, C.B. Carter, U. Kortshagen & S.A. Campell, 2005. Single nanoparticle semiconductor devices. IEEE Trans. Electron Dev. (accepted for publication).

Ding Z., B.M. Quinn, S.K. Haram, L.E. Pell, B.A. Korgel & A.J. Bard, 2002. Electrochemistry and electrogenerated chemiluminescence from silicon nanocrystal quantum dots. Science 296, 1293–1297.

Draeger E.W., J.C. Grossman, A.J. Williamson & G. Galli, 2004. Optical properties of passivated silicon nanoclusters: The role of synthesis. J. Chem. Phys. 120(22), 10807–10814.

Eaglesham D.J., A.E. White, L.C. Feldman, N. Moriya & D.C. Jacobson, 1993. Equilibrium shape of Si. Phys. Rev. Lett. 70(11), 1643–1646.

Ehbrecht M. & F. Huisken, 1999. Gas-phase characterization of silicon nanoclusters produced by laser pyrolysis of silane. Phys. Rev. B: Condens. Matter Mater. Phys. 59(4), 2975–2985.

Franzò G., A. Irrera, E.C. Moreira, M. Miritello, F. Iacona, D. Sanfilippo, G. Di Stefano, P.G. Fallica & F. Priolo, 2002. Electroluminescence of silicon nanocrystals in MOS structures. Appl. Phys. A: Mat. Sci. Proc. 74, 1–5.

Friedlander S.K., 2000 Smoke, Dust, and Haze – Fundamentals of Aerosol Dynamics. New York: Oxford University Press.

Fu Y., M. Willander, A. Dutta & S. Oda, 2000. Carrier conduction in a Si-nanocrystal-based single-electron transistor-I. Effect of gate bias. Superlattices Microstruct 28(3), 177–187.

Furukawa S. & T. Miyasato, 1988. Three-dimensional quantum well effects in ultrafine silicon particles. Jpn. J. Appl. Phys. 27(11), L2207.

Gerberich W.W., W.M. Mook, C.R. Perrey, C.B. Carter, M.I. Baskes, R. Mukherjee, A. Gidwani, J. Heberlein, P.H. McMurry & S.L. Girshick, 2003. Superhard silicon nanospheres. J. Mech. Phys. Solids. 51, 979–992.

Goldstein A.N., C.M. Echer & A.P. Alivisatos, 1992. Melting in semiconductor nanocrystals. Science 256, 1425–1427.

Goree J., 1994. Charging of particles in a plasma. Plasma Sources Sci. Technol. 3, 400.

Holmes J.D., K.J. Ziegler, C. Doty, L.E. Pell, K.P. Johnston & B.A. Korgel, 2001. Highly luminescent silicon nanocrystals with discrete optical transitions. J. Am. Chem. Soc. 123, 3743–3748.

Holtz R.L., V. Provenzano & M.A. Imam, 1996. Overview of nanophase metals and alloys for gas sensors, getters, and hydrogen storage. Nanostruct. Mater. 7, 259–264.

Huisken F., D. Amans, G. Ledoux, H. Hofmeister, F. Cichos & J. Martin, 2003. Nanostructuration with visible-light-emitting silicon nanocrystals. New J. Phys. 5, 1–10Paper No. 10.

Kennedy M.K., F.E. Kruis, H. Fissan & B.R. Mehta, 2003. Fully automated, gas sensing, and electronic parameter measurement setup for miniaturized nanoparticle gas sensors. Rev. Sci. Instr. 74(11), 4908–4915.

Kennedy M.K., F.E. Kruis, H. Fissan, B.R. Mehta, S. Stappert & G. Dumpich, 2003. Tailored nanoparticle films from monosized tin oxide nanocrystals: particle synthesis, film formation, and size-dependent gas-sensing properties. J. Appl. Phys. 93(1), 551–560.

Kim T.W., D.C. Choo, J.H. Shim & S.O. Kang, 2002. Single-electron transistors operating at room temperature, fabricated utilizing nanocrystals created by focused-ion beam. Appl. Phys. Lett. 80(12), 2168–2170.

Klein D.L., R. Roth, A.K.L. Lim, A.P. Alivisatos & P.L. McEuen, 1997. A single-electron transistor made from a cadmium selenide nanocrystal. Nature 389, 699–701.

Kortshagen U. & U. Bhandarkar, 1999. Modeling of particulate coagulation in low pressure plasmas. Phys. Rev. E 60(1), 887.

Ledoux G., J. Gong, F. Huisken, O. Guillois & C. Reynaud, 2002. Photoluminescence of size-separated silicon nanocrystals: confirmation of quantum confinement. Appl. Phys. Lett. 80(25), 4834–4836.

Ledoux G., O. Guillois, D. Porterat, C. Reynaud, F. Huisken, B. Kohn & V. Paillard, 2000. Photoluminescence properties of silicon nanocrystals as a function of their size. Phys. Rev. B 62(23), 15942–15951.

Li X., Y. He, S.S. Talukdar & M.T. Swihart, 2003. Process for preparing macroscopic quantities of brightly photoluminescent silicon nanoparticles with emission spanning the visible spectrum. Langmuir 19(20), 8490–8496.

Littau K.A., P.J. Szajowski, A.J. Muller, A.R. Kortan & L.E. Brus, 1993. A luminescent silicon nanocrystal colloid via a high-temperature aerosol reaction. J. Phys. Chem. 97, 1224–1230.

Mangolini L., E. Thimsen & U. Kortshagen, 2005. High-yield plasma synthesis of luminescent silicon nanocrystals. Nano Lett. 5(4), 655–659.

Matsoukas T., 1997. The coagulation rate of charged aerosols in ionized gases. J. Colloid. Interface Sci. 187, 474.

Matsoukas T. & M. Russel, 1995. Particle charging in low-pressure plasmas. J. Appl. Phys. 77, 4285.

Nayfeh M., O. Akcakir, J. Therrien, Z. Yamani, N. Barry, W. Yu & E. Gratton, 1999. Highly nonlinear photoluminescence threshold in porous silicon. Appl. Phys. Lett. 75(26), 4112–4114.

Nayfeh M.H., N. Barry, J. Therrien, O. Akcakir, E. Gratton & G. Belomoin, 2001. Stimulated blue emission in reconstituted films of ultrasmall silicon nanoparticles. Appl. Phys. Lett. 78(8), 1131–1133.

Nishiguchi K. & S. Oda, 2000. Electron transport in a single silicon quantum structure using a vertical silicon probe. J. Appl. Phys. 88(7), 4186–4190.

Onischuk A.A., A.I. Levykin, V.P. Strunin, K.K. Sabelfeld & V.N. Panfilov, 2000. Aggregate formation under homogeneous silane thermal decomposition. J. Aerosol Sci. 31(11), 1263–1281.

Onischuk A.A., A.I. Levykin, V.P. Strunin, M.A. Ushakova, R.I. Samoilova, K.K. Sabelfeld & V.N. Panfilov, 2000. Aerosol formation under heterogeneous/homogeneous thermal decomposition of silane: experiment and numerical modeling. J. Aerosol Sci. 31(8), 879–906.

O'Regan B. & M. Grätzel, 1991. A low-cost, high-efficiency solar cell based on dye-sensitized colloidal TiO$_2$ films. Nature 353(6346), 737.

Ostraat M.L., J.W. De Blauwe, M.L. Green, L.D. Bell, H.A. Atwater & R.C. Flagan, 2001a. Ultraclean two-stage aerosol reactor for production of oxide-passivated silicon nanoparticles for novel memory devices. J. Electrochem. Soc. 148(5), G265–G270.

Ostraat M.L., J.W. De Blauwe, M.L. Green, L.D. Bell, M.L. Brongersma, J. Casperson, R.C. Flagan & H.A. Atwater, 2001b. Synthesis and characterization of aerosol silicon nanocrystal nonvolatile floating-gate memory devices. Appl. Phys. Lett. 79(3), 433–435.

Park N.-M., T.-S. Kim & S.-J. Park, 2001. Band gap engineering of amorphous silicon quantum dots for light-emitting diodes. Appl. Phys. Lett. 78(17), 2575–2577.

Pettigrew K.A., Q. Liu, P.P. Power & S.M. Kauzlarich, 2003. Solution synthesis of alkyl- and alkyl/alkoxy-capped silicon nanoparticles via oxidation of Mg$_2$Si. Chem. Mater. 15(21), 4005–4011.

Puzder A., A.J. Williamson, J.C. Grossman & G. Galli, 2002. Surface chemistry of silicon nanoclusters. Phys. Rev. Lett. 88(9), 097401–097404.

Puzder A., A.J. Williamson, J.C. Grossman & G. Galli, 2003. Computational studies of the optical emission of silicon nanocrystals. J. Am. Chem. Soc. 125(9), 2786–2791.

Reboredo F.A., A. Franceschetti & A. Zunger, 1999. Excitonic transitions and exchange splitting in Si quantum dots. Appl. Phys. Lett. 75(19), 2972–2974.

Sankaran R.M., D. Holunga, R.C. Flagan & K.P. Giapis, 2005. Synthesis of blue luminescent Si nanoparticles using atmospheric-pressure microdischarges. Nano Lett. 5(3), 531–535.

Schweigert V.A. & I.V. Schweigert, 1996. Coagulation in low-temperature plasmas. J. Phys. D: Appl. Phys. 29, 655.

Selwyn G.S., K.L. Haller & E.F. Patterson, 1993. Trapping and behavior of particulates in a radio frequency magnetron etching tool. J. Vac. Sci. Technol. A 11, 1132.

Selwyn G.S., J.E. Heidenreich & H.L. Haller, 1990. Particle trapping phenomena in radio frequency plasmas. Appl. Phys. Lett. 57, 1876.

Shen Z., T. Kim, U. Kortshagen, P. McMurry & S. Campbell, 2003. Formation of highly uniform silicon nanoparticles in high density silane plasmas. J. Appl. Phys. 94(4), 2277–2283.

Shi F.G., 1994. Size dependent thermal vibrations and melting in nanocrystals. J. Mater. Res. 9(5), 1307–1312.

St. John J., J.L. Coffer, Y. Chen & R.F. Pinizzotto, 1999. Synthesis and characterization of discrete luminescent erbium-doped silicon nanocrystals. J. Am. Chem. Soc. 121, 1888–1892.

Stekolnikov A.A., J. Furthmüller & F. Bechstedt, 2002. Absolute surface energies of group-IV semiconductors: dependence on orientation and reconstruction. Phys. Rev. B 65, 115318.

Stoffels E., W.W. Stoffels, G.M.W. Kroesen & F.J.D. Hoog, 1996. Dust formation and charging in an Ar/SiH_4 radio-frequency discharge. J. Vac. Sci. Technol. A 14, 556.

Takahashi N., H. Ishikuro & T. Hiramoto, 2000. Control of Coulomb blockade oscillations in silicon single electron transistors using silicon nanocrystal floating gates. Appl. Phys. Lett. 76(2), 209–211.

Tiwari S., F. Rana, K. Chan, L. Shi & H. Hanafi, 1996a. Single charge and confinement effects in nano-crystal memories. Appl. Phys. Lett. 69, 1232.

Tiwari S., F. Rana, H. Hanafi, A. Hartstein, E.F. Crabbé & K. Chan, 1996b. A silicon nanocrystals based memory. Appl. Phys. Lett. 68, 1377.

Vasiliev I., J.R. Chelikowsky & R.M. Martin, 2002. Surface oxidation effects on the optical properties of silicon nanocrystals. Phys. Rev. B (Condensed Matter and Materials Physics) 65(12), 121302.

Volkening F.A., M.N. Naidoo, G.A. Candela, R.L. Holtz & V. Provenzano, 1995. Characterization of nanocrystalline palladium for solid state gas sensor applications. Nanostruct. Mater. 5, 373–382.

Watanabe Y. & M. Shiratani, 1993. Growth kinetics and behavior of dust particles in silane plasmas. Jpn. J. Appl. Phys. 32, 3074.

Wilcoxon J.P. & G.A. Samara, 1999. Tailorable, visible light emission from silicon nanocrystals. Appl. Phys. Lett. 74(21), 3164–3166.

Wilcoxon J.P., G.A. Samara & P.N. Provencio, 1999. Optical and electronic properties of Si nanoclusters synthesized in inverse micelles. Phys. Rev. B 60(4), 2704–2714.

Wolkin M.V., J. Jorne, P.M. Fauchet, G. Allan & C. Delerue, 1999. Electronic states and luminescence in porous silicon quantum dots: the role of oxygen. Phys. Rev. Lett. 82(1), 197.

Zhang Z., M. Zhao & Q. Jiang, 2001. Melting temperature of semiconductor nanocrystals in the mesoscopic size range. Semicond. Sci. Technol. 16, L33–L35.

Zhou Z., L. Brus & R. Friesner, 2003a. Electronic structure and luminescence of 1.1- and 1.4-nm silicon nanocrystals: oxide shell versus hydrogen passivation. Nano Lett. 3(2), 163–167.

Zhou Z., R.A. Friesner & L. Brus, 2003b. Electronic structure of 1 to 2 nm diameter silicon core/shell nanocrystals: surface chemistry, optical spectra, charge transfer, and doping. J. Am. Chem. Soc. 125, 15599–15607.

Journal of Nanoparticle Research (2007) 9:53–59
DOI 10.1007/s11051-006-9156-8

© Springer 2006

Special focus: Nanoparticles and Occupational Health

Rationale and principle of an instrument measuring lung deposited nanoparticle surface area

H. Fissan*, S. Neumann, A. Trampe, D.Y.H. Pui and W.G. Shin
*Universität Duisburg-Essen, Ingenieurwissenschaften, Nanostrukturtechnik, Bismarckstr. 81, D-47048, Duisburg, Germany; *Author for correspondence (Tel.: +49-203-3793201; Fax: +49-203-3793268; E-mail: heinz. fissan@uni-due.de)*

Received 27 July 2006; accepted in revised form 2 August 2006

Key words: nanotechnology, nanoparticles, aerosol, health risk, exposure control, exposure measurement, lung deposited surface area, lung simulator, Electrical Aerosol Detector (EAD), Occupational health

Abstract

The risk of nanoparticles by inhalation for human health is still being debated but some evidences of risk on specific properties of particles < 100 nm diameter exist. One of the nanoparticle parameters discussed by toxicologists is their surface area concentration as a relevant property for e.g. causing inflammation. Concentrations of these small particles ($\sim$ < 100 nm) are currently not measured, since the mass concentrations of these small particles are normally low despite large surface area concentrations. Airborne particles will always be polydisperse and show a size distribution. Size is normally described by an equivalent diameter to include deviations in properties from ideal spherical particles. Here only nanoparticles below a certain size to be defined are of interest. Total concentration measures are determined by integration over the size range of interest. The ideal instrument should measure the particles according to the size weighting of the wanted quantity. Besides for the geometric surface area the wanted response function can be derived for the lung deposited surface area in the alveolar region. This can be obtained by weighting the geometric surface area as a function of particle size with the deposition efficiency for the alveolar region for e.g. a reference worker for work place exposure determination. The investigation of the performance of an Electrical Aerosol Detector (EAD) for nearly spherical particles showed that its response function is close to the lung deposited surface areas in different regions of the human respiratory system. By changing the ion trap voltage an even better agreement has been achieved. By determining the size dependent response of the instrument as a function of ion trap voltage the operating parameters can be optimized to give the smallest error possible. Since the concept of the instrument is based on spherical particles and idealized lung deposition curves have been used, in all other cases errors will occur, which still have to be defined. A method is now available which allows in principle the determination of the total deposited surface area in different regions of the lung in real time. It can easily be changed from one deposited region to another by varying the ion trap voltage. It has the potential to become a routine measurement technique for area measurements and personal control in e.g. work place environments.

Introduction

Nanotechnology offers great opportunities for new and improved nanostructured, functionalized materials and devices. Besides thin films the most important building blocks are nanoparticles (Kruis et al., 1998), which can be produced in a solid, liquid or gaseous matrix. Product nanoparticles are mostly defined to be smaller than 100 nm and down to a few nanometers. The upper size limit is depending on the problem raised and will be defined in this paper for the case of measuring lung deposited nanoparticle surface area. To make use of some of these properties, nanoparticles have to be single isolated particles. In other cases they may consist of aggregated and agglomerated primary nanoparticles.

The mass production of nanoparticles in the gas phase has several advantages, because clean and continuous processing is possible. On the other hand nanoparticles in the gas phase have a high mobility. There are increasing chances for them to escape during gas phase processing, handling and use, than in liquid processes. The particles emitted to work places or more generally into the environment are easily transported with the gas flow to human lungs. They are inhaled and deposited in the nose or mouth and in different parts of the lung. They either may cause negative health effects at the point of deposition or may be transmitted to other end organs (Kreyling et al., 2002; Oberdörster et al., 1995).

The risk of nanoparticle intake is dependent on exposure and hazard. Here we are interested in exposure measurement. For this purpose, monitors for exposure control are needed in the work place, controlling either an area or a person at the point of possible nanoparticle intake. We are introducing a new concept based on an existing instrument for area monitoring, which allows the measurements of the nanoparticle surface area deposited in different parts of the human respiratory system.

Nanoparticle surface area measurement

The measurement objects of current interest are nanoparticles (< 100 nm). Most standards set up by different organizations all over the world for work place measurement thus far are based on mass concentration limits. Mass measurement methods are not sufficiently sensitive for airborne nanoparticles and may not be sensitive toward the specific health relevant properties of nanoparticles. The most sensitive concentration measure in this particle range (< 100 nm diameter) is the number concentration. Unfortunately the number concentration is dominated by very small particles, which

are difficult to measure because of increasing line losses and decreasing counting efficiency with decreasing particle size for all counters.

On the other hand the most important question, which has yet to be raised, is whether the number concentration correlates with health effects. This is true for asbestos fibers with a certain probability for each fiber to cause a negative health effect and may be also for nanoparticles in case of clogging after penetrating into the blood. For nanoparticles, toxicologists discuss the particle surface area among others as a relevant measure (Oberdörster, 1996; Donaldson et al., 1998), because most of the processes in the human body environment take place via the particle surface, which is increasing significantly with decreasing particle size in the nanometer size range for the same amount of mass. The health effects after intake are strongly depending also on the deposition regions. Particularly discussed are the deposition in the nose (head), because of possible transfer of nanoparticles to the brain, the tracheobronchial region as well as the alveolar region, because of inefficiency of clearing mechanism and the possible transfer to the blood circulation system with resulting distribution in several end organs (Kreyling et al., 2002).

In Figure 1 the deposition curves for head (H), tracheobronchial (TB) and alveolar (A) deposition are shown. They were obtained using the UK National Radiological Protection Board's (NRPB's) LUDEP Software (James et al., 2000), based on the recommendations of ICRP Publication 66 (ICRP, 1994). Different people performing different activities have different deposition curves. We have chosen a reference worker with the following conditions:

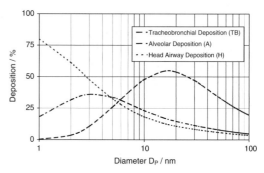

Figure 1. Deposition curves (James et al., 2000).

– Breathing type: nose only
– Functional residual capacity: 3301 cc
– Breathing rate (Breath/min): 20
– Ventilation rate: 1.5 m³/h
– Activity level: light exercise

From all these considerations it follows that an instrument is needed, which is capable of measuring the total nanoparticle surface area fractions originating from nanoparticle processing, which are deposited in different parts of the human respiratory system.

Needed instrument response

To be able to perform on-line measurements and easy data evaluation the instrument should deliver an electrical signal. If possible, the wanted instrument should have a linear response to particle surface area. Using monodisperse, spherical particles of known size the instrument then can be calibrated by performing parallel number concentration measurements. Knowing number concentration and particle size the surface can easily be calculated. For each size a calibration factor can be determined. Since the calibration is based on electrical mobility diameter of spheres for sizing and the unipolar charging for the generation of the electrical signal, only an equivalent surface concentration based on these processes can be determined in case of agglomerates.

Most emitted particles are distributed as a function of size. If we want to measure the total concentration of a polydisperse aerosol (total number, -surface area, -mass or other -weighted quantities), the instrument has to show a certain size dependent sensitivity depending on the size weighting of the wanted quantity. This sensitivity can be derived by describing the integral over the size distribution by its sum. Each size increment in the sum is proportional to the number concentration times the corresponding diameter weighting, in case of geometric surface area, D_p^2. To normalize the size dependant response with respect to the number concentration, 100 nm particles have been chosen as a reference and their normalized sensitivity has been set to be 1. The needed sensitivities as function of particle size are shown in Figure 2. The surface area shows a D_p^2-dependency and the number is independent of particle size (D_p^0). If we now weight the sensitivities for

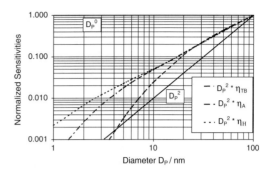

Figure 2. Response function for lung deposited surface area in comparison with response function for number concentration (D_p^0) and geometric surface area (D_p^2).

geometric surface area with the corresponding deposition curves (deposition efficiencies η, see Figure 1), we simulate the deposition in the different regions of the human respiratory system. In Figure 2 the needed response functions for head, tracheobronchial and alveolar depositions are shown. They are nonlinear and less steep compared with the response function for geometric surface area, but further away from the constant curve for number concentration.

Instruments based on diffusion charging and their response functions

Diffusion charging is a process which is at least in certain particle size regions proportional to particle surface (Rogak et al., 1993; Jung & Kittelson, 2005). Depending on instrument design the sensitivity as a function of particle size, i.e. the response function, is different because of differences in the charging process and particle losses. Ku and Maynard (2005) have recently shown that the Matter instrument, Switzerland, e.g., the instrument LQ1-DC, does show a D_p^2-dependency in the size range between 30 and 100 nm. Above 100 nm and below 30 nm the sensitivities are smaller then the one needed for a D_p^2-dependency. For these studies monodisperse silver agglomerates were used, synthesized by evaporating silver in a ceramic boat in a furnace. In the cold carrier gas (nitrogen) silver particles were formed, which agglomerated rapidly. In a second oven they were heat treated (Kruis & Fissan, 1999). The polydisperse aerosol was size fractionated using a Differential Mobility Analyzer (DMA). With a second

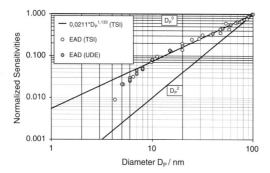

Figure 3. Response function of EAD 3070A.

electrometer to measure the current caused by the deposited charges. The response function of the instrument changes with the applied voltage to the ion trap, because with increasing voltage particles are also eliminated. The instrument was challenged with differently sized, monodisperse silver nanoparticles in the size range between 10 and 100 nm with an ion trap voltage between 20 and 200 V. The result is shown in Figure 4. The normalized sensitivity is 1.0 for the reference case of 100 nm particles. For all voltages it is decreasing with decreasing particle size, because of a reduction of the average charge level. With decreasing particle size and increasing ion trap voltage the particle losses in the trap also increase causing a further reduction in normalized sensitivity. In Figure 5 the response functions for different ion trap voltages derived from the data in Figure 4 are shown as function of particle size for better comparison with the response functions shown earlier. Especially in the small size range the normalized sensitivity drops more steeply with

furnace the particles were sintered at increasing temperature, which did not lead to significant differences in the response function. This demonstrates that the sensitivity is not a strong function of the shape of the particles. They can be assumed to behave like spherical silver particles with the same electrical mobility as the agglomerates.

We performed similar experiments with the Electrical Aerosol Detector (EAD)/TSI 3070A, using unsintered silver agglomerates (see Figure 3). Between 10 nm and 100 nm the normalized sensitivity can be described by the function $0.0211 \cdot D_p^{1.133}$. Below 10 nm the response function drops more sharply compared with the given function. Our data compare well with earlier data given by the manufacturer (Kaufman et al., 2002). If we compare this function with the weighted response functions for lung deposited surface areas (see Figure 2) they match well in the size range 20–100 nm. This brought up the idea to manipulate the EAD response so that it matches better the needed responses for different weightings of lung deposition.

Modification of EAD to measure lung deposited surface area

The EAD consists of a charging chamber where the aerosol is mixed with positively charged ions, which attach to the particles by diffusion. The unipolarly charged aerosol with residual ions is then introduced into an ion trap to which a voltage of 20 V is applied. The highly mobile residual ions are eliminated in the electric field. The charged particles are then collected in a filter downstream of the ion trap which is part of a sensitive

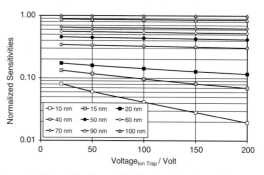

Figure 4. Normalized sensitivities as function of ion trap voltage for different sized particles.

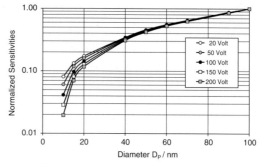

Figure 5. Response functions for different ion trap voltages.

increasing voltage. With increasing voltage larger amounts of small particles are eliminated, causing an increasing drop in normalized sensitivity. Major changes occur only below 40 nm. By eye fitting we have chosen the response function for 100 V ion trap voltage to be a close fit to the needed response function for tracheobronchial deposition. In Figure 6 the measured response function for 100 V ion trap voltage is compared with the response function for tracheobronchial deposition. A deviation occurs only for the very small (< 10 nm) particles. For 200 V a good comparison is achieved for the alveolar deposition (Figure 7). The differences between the normalized response functions for alveolar and trachiobronchial deposition are rather small. The main difference is caused by the difference in deposition efficiency at the reference point. This is taken into account through the calibration, which includes the different deposition efficiencies at the reference point of 100 nm particle size. The described

procedure for determining the normalized sensitivity as function of particle size allows also the determination of the calibration curves. Multiplying the normalized sensitivities with different number concentrations allows the determination of the lung deposited surface area as function of electrometer current. As long as the response function of the instrument is equal to the needed response function for different particle sizes the data from all measurements can be taken to construct the calibration curve (Figure 8), since they all lead to the same calibration factor. The calibration curves are-linear in the covered concentration range and can be described by the given simple functions. In Figure 9 the calibration factors are plotted as function of particle size. The dotted lines refer to an exceptable error of ± 25%. In both cases, tracheobronchial and alveolar deposition, the calibration factors are outside of this range only below 10 nm. Fortunately the error contribution to the total surface area of a

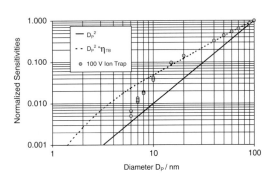

Figure 6. Response function of EAD 3070A – 100 V ion trap.

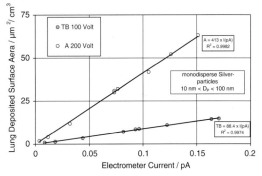

Figure 8. Calibration curve for EAD for tracheobronchial and alveolar lung area (Lung Simulator).

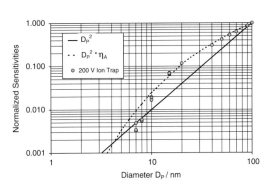

Figure 7. Response function of EAD 3070A – 200 V ion trap.

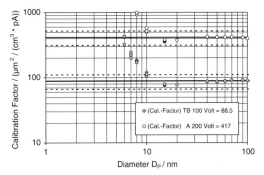

Figure 9. Calibration factor as function of particle size for 100 and 200 V.

polydisperse aerosol by smaller particles is *negligible* because of the D_p^2-dependency. The surface area contribution of a 10 nm particle is only 1% of that of a 100 nm particle.

Up to now we assume that the instrument cuts out all particles larger than 100 nm. This may not be possible, causing large errors. Another upper limit could be the minimum in the deposition curves around 300 nm to include all deposited small particles. Above ~400 nm the deposition curves become density dependent and therefore material dependent.

In any case it still has to be shown that the response function of the instrument in the size range between 100 nm and any larger size follows the needed response function.

Optimal ion-trap voltage

Under certain conditions the available performance data (Figure 4) can be used to choose optimal ion-trap voltages for which the error is minimal. The absolute error in the measured total deposited surface area is depending on the actual size distribution, which we don't know *a priori*. We assume a constant number distribution over the particle size range between 10 and 100 nm. The real size distribution of nanoparticles will normally show smaller number concentrations at both ends of the covered size range between 10 and 100 nm compared with the maximal concentration. This leads to an overestimation of the error at both ends of the size range. The error contribution of the small particles is small, because of the D_p^2 dependency. At the upper end of the size distribution the influence on the sum of errors becomes more important.

We can interpolate or calculate the fitted normalized sensitivities for each ion-trap voltage between 20 V and 200 V for measured particle diameters. The normalized sensitivities for the wanted response function are calculated as described earlier. The squared differences between the calculated normalized sensitivities for a given ion-trap voltage and wanted sensitivity at measured particle diameter are calculated and summed up. Figure 10 shows the results for tracheobronchial, alveolar and head airways deposition. The minimum of these curves of the sum of the errors in case of a constant size distribution within the size range of 10–100 nm represent the optimal ion-trap voltages, where both response functions show the best agreement.

Table 1 shows the optimal voltages and least square sums for the different deposition areas. Also the calibration factor has been derived. For comparison the applied voltages and their corresponding data are also shown. The differences in least square sum as well as in calibration factor are not very large between the different deposition areas. This is due to the fact that the major differences in the response functions occur below

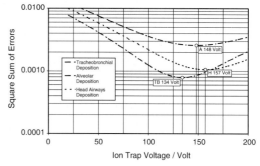

Figure 10. Determination of optimal ion trap voltage.

Table 1. Optimal ion trap voltages

Lung deposition	Measured ion-trap voltage	Optimal ion-trap voltage	Least square sum	Calibration factor
Tracheobronchial deposition	20 V		0.00774	83.3 μm²/(cm³ *pA)
	100 V		0.00125	89.7 μm²/(cm³ *pA)
		134 V	0.00079	92.5 μm²/(cm³ *pA)
Alveolar deposition	20 V		0.01213	347 μm²/(cm³ *pA)
		148 V	0.00259	391 μm²/(cm³ *pA)
	200 V		0.00353	409 μm²/(cm³ *pA)
Head airway deposition	20 V		0.00987	62.4 μm²/(cm³ *pA)
		157 V	0.00106	70.7 μm²/(cm³ *pA)

10 nm, a size range, which does not contribute much to the total surface of a polydisperse aerosol, even if we consider only particles below 100 nm.

Summary

Modifications of the EAD with different ion trap voltages have been tested with the goal of determining the deposited nanoparticle surface area for different regions of the human respiratory system. We have demonstrated that this can be achieved for nanoparticles below 100 nm. The investigation was performed using agglomerated silver nanoparticles. Further studies need to be conducted to investigate what influence different particle shapes and materials may have. Also further studies on the possibilities of modifying the response to get better agreement for different wanted response functions should be investigated. For reference cases additional quantities like dose, dose per lung area or lung mass may easily be derived.

References

Donaldson K., X.Y. Li & W. MacNee, 1998. Ultrafine (nanometer) particle mediated lung injury. J. Aerosol Sci. 29(5–6), 553–560.

ICRP., 1994. International Commission on Radiological Protection Publication 66 Human Respiratory Tract Model for Radiological Protection. Oxford, Pergamon: Elsevier Science Ltd.

James A.C., M.R. Bailey & M-D. Dorrian, 2000. LUDEP Software, Version 2.07: Program for implementing ICRP-66 Respiratory tract model. RPB, Chilton, Didcot, OXON. OX11 ORQ UK.

Jung H. & D. Kittelson, 2005. Measurement of electrical charge on diesel particles. Aerosol Sci. Technol. 39(12), 1129–1135.

Kaufman S.L., A. Medved, A. Pöcher, N. Hill, R. Caldow & F.R. Quant, 2002. An electrical aerosol detector based on the corona-jet charger. AAAR conference (poster).

Kreyling W.G., M. Semmler, F. Erbe, P. Mayer, S. Takenaka, H. Schulz, G. Oberdörster & A. Ziesenis, 2002. Translocation of ultrafine insoluble iridium particles from lung epithelium to extrapulmonary organs is size dependent but very low. J. Toxicol. Environ. Health Pt A 65(20), 1513–1530.

Kruis F.E., H. Fissan & A. Peled, 1998. Synthesis of nanoparticles in the gas phase for electronic, optical and magnetic application – A review. J. Aerosol Sci. 29(5–6), 511–535.

Kruis F.E. & H. Fissan, 1999. Nano-Process Technology for Synthesis and Handling of Nanoparticles. Kona Powder and Particles, No.17, pp. 130–139.

Ku B.K. & A.D. Maynard, 2005. Generation and investigation of airborne Ag nanoparticles with specific size and morphology by homogeneous nucleation, coagulation and sintering. J. Aerosol Sci. 36(9), 1108–1124.

Oberdörster G., R.M. Gelein, J. Ferin & B. Weiss, 1995. Association of particulate air pollution and acute mortality: Involvement of ultrafine particles. Inhal. Toxicol. 7, 111–124.

Oberdörster G., 1996. Significance of particle parameters in the evaluation of exposure-dose–response relationships of inhaled particles. Particul. Sci. Technol. 14(2), 135–151.

Rogak S.N., R.C. Flagan & H.V. Nguyen, 1993. The mobility and structure of aerosol agglomerates. Aerosol Sci. Technol. 18(1), 25–47.

Journal of Nanoparticle Research (2007) 9:61–69
DOI 10.1007/s11051-006-9153-y

Special focus: Nanoparticles and Occupational Health

Calibration and numerical simulation of Nanoparticle Surface Area Monitor (TSI Model 3550 NSAM)

W.G. Shin[1], D.Y.H. Pui[1,*], H. Fissan[2], S. Neumann[2] and A. Trampe[2]
*[1]Mechanical Engineering Department, University of Minnesota, Minneapolis, MN, USA; [2]University of Duisburg-Essen, Duisburg, Germany; *Author for correspondence (E-mail: dyhpui@tc.umn.edu)*

Received 26 July 2006; accepted in revised form 1 August 2006

Key words: nanoparticle surface area, deposition in compartments of human lung, tracheobronchial, alveolar, instrumentation, occupational health

Abstract

TSI Nanoparticle Surface Area Monitor (NSAM) Model 3550 has been developed to measure the nanoparticle surface area deposited in different regions of the human lung. It makes use of an adjustable ion trap voltage to match the total surface area of particles, which are below 100 nm, deposited in tracheobronchial (TB) or alveolar (A) regions of the human lung. In this paper, calibration factors of NSAM were experimentally determined for particles of different materials. Tests were performed using monodisperse (Ag agglomerates and NaCl, 7–100 nm) and polydisperse particles (Ag agglomerates, number count mean diameter below 50 nm). Experimental data show that the currents in NSAM have a linear relation with a function of the total deposited nanoparticle surface area for the different compartments of the lung. No significant dependency of the calibration factors on particle materials and morphology was observed. Monodisperse nanoparticles in the size range where the response function is in the desirable range can be used for calibration. Calibration factors of monodisperse and polydisperse Ag particle agglomerates are in good agreement with each other, which indicates that polydisperse nanoparticles can be used to determine calibration factors. Using a CFD computer code (Fluent) numerical simulations of fluid flow and particle trajectories inside NSAM were performed to estimate response function of NSAM for different ion trap voltages. The numerical simulation results agreed well with experimental results.

Introduction

Occupational health risks associated with manufacturing and application of nanoparticles is one of the critical issues for nanotechnology development. Demands for nanomaterials are rapidly increasing, and workers may be exposed to particularly harmful nanoparticles. Several recent studies have shown that the toxicity of inhaled nanoparticles may be more appropriately associated with particulate surface area (Oberdorster et al., 1995; Oberdorster, 1996, 2005; Donaldson et al., 1998; Maynard & Kuempel, 2005). Nanoparticles defined to be below 100 nm have an increasing surface area with a decreasing particle size for the same amount of mass. From the viewpoint of nanoparticle toxicity, an instrument which measures nanoparticle surface area deposited in the human lung is very desirable.

Relatively few techniques are available to monitor exposures with respect to aerosol surface area

62

(Shi et al., 2001; Maynard, 2003; Jung & Kittelson, 2005). The BET method based on a gas adsorption method is not suited for a rapid evaluation of aerosol surface area at lower concentration (Brunauer et al., 1938). It can be used only for powders, not for particles in the gasborne state. It does not have on-line capabilities. The first instrument designed to measure aerosol surface-area was the epiphaniometer (Baltensperger et al., 1988). The epiphaniometer is not well suited for widespread use at the workplace because of the inclusion of radioactive source and the lack of effective temporal resolution. One of other possible methods is diffusion charging (DC). Instruments using DC include the LQ1-DC diffusion charger (Matter Engineering, Switzerland) and the TSI model 3070a Electrical Aerosol Detector (EAD). These diffusion chargers were recently evaluated (Jung & Kittelson, 2005; Ku & Maynard, 2005). Several studies using atmospheric field data have shown that EAD can be used as a useful indicator of the amount of particle surface area deposited in the lung (Woo et al., 2001; Wilson et al., 2003, 2004). Recently, it was found that response functions of EAD can be changed to match the particle surface area deposited in the lung through the adjustment of ion trap voltage in EAD (Fissan & Kuhlbusch, 2005). Based on the observation, TSI Nanoparticle Surface Area Monitor (NSAM) model 3550 has been developed to measure the nanoparticle

surface area deposited in two regions, trancheobronchial (TB) and alveolar (A), of the human lung of a reference worker (Fissan et al., 2006) by adjusting ion trap voltage. Errors will occur in other cases such as kids and asthmatics; further studies are still needed for other cases. NSAM provides a simple and fast solution for measuring the surface area dose in different parts of the inhalation system.

The schematic of NSAM is shown in Figure 1. The major components of NSAM consist of a diffusion charger chamber, an ion trap, an electrometer filter, and other sharp bending parts to transport the nanoparticle stream. The total inlet flow rate of 2.5 lpm is divided into 1.5 lpm for aerosol and 1.0 lpm for sheath air surrounding a corona needle. In the charger chamber, the aerosol stream and the ion stream opposing each other are mixed. After the aerosols are charged by unipolar ions in the charger chamber, the flow with a flow rate of 2.5 lpm enters the ion trap. Excess particles with high electrical mobility and ions are removed in the ion trap. Particles which penetrate the ion trap are collected by the electrometer filter. The delivered charges give rise to a current, which is measured by an electronic circuit.

Modifications of the EAD with different ion trap voltages have been tested with the goal of determining the deposited nanoparticle surface area for different regions of the human inhalation

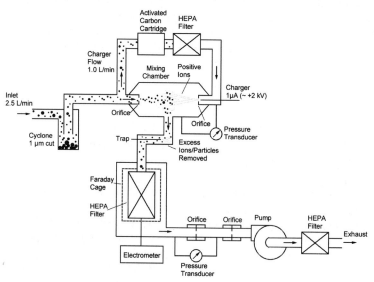

Figure 1. The schematic of NSAM.

63

system (Fissan & Kuhlbusch, 2005, Fissan et al., 2006). The response function curve of EAD for an ion trap voltage of 100 V matched with TB deposition and an ion trap voltage of 200 V with A deposition. The EAD with variable ion trap voltages to match the lung deposition curves is now referred as NSAM.

This paper consists of two parts. In the first part, response functions and calibration factors of NSAM for TB and A regions are experimentally obtained. The effects of particle materials and shape on the response function and calibration factor of NSAM are investigated. Tests using polydisperse and monodisperse Ag particle agglomerates are compared to each other. In the second part, numerical simulation results for the response function of NSAM are compared with experimental data.

Experimental methods

In order to obtain both the response function and the calibration curve of NSAM in which the input is the particle number concentration (N) and the output is the electrometer current (I), a test facility was constructed to generate Ag particle agglomerates with an electrical mobility size range from 7 to 100 nm (Figure 2). Ag wire (purity level 99.9 %) was placed in a ceramic boat in the furnace. Nitrogen with a flow rate of 3.0 lpm was used as a carrier gas passing through the electric furnace. The flow rate of the carrier gas was regulated by

combinations of needle valve, pressure gauge, and rotameter. Silver was vaporized in the electric furnace followed by particle formation by condensation and coagulation (Han et al., 2000). Different experimental setups were added following the electric furnace depending on monodisperse or polydisperse test. In the monodisperse test, Ag particles are introduced to both an Ultrafine Condensation Particle Counter (UCPC) and NASM after primary particles are classified by a Differential Mobility Analyzer (DMA) set at a fixed voltage corresponding to particle size. In the polydisperse test, a scanning mobility particle sizer (SMPS) and UCPC are used to measure particle size distribution and polydisperse Ag particles are directly introduced to NSAM after passing through a neutralizer. For all experiments, the carrier gas flow rate was fixed at 3.0 lpm. In order to change particle size distributions, the electric furnace temperature was varied from 950 to 1200 °C. The dilution flow using compressed nitrogen gas (1.0 lpm) was also introduced in front of UCPC and NSAM. Aerosol flow after the dilution was split into two streams going into UCPC (1.5 lpm) and NSAM (2.5 lpm), respectively. For the generation of monodisperse particles, primary particles larger than the peak size of a given particle size distribution were subsequently selected with a DMA (Model 3080, TSI, Inc.). A Kr-85 neutralizer was used to get a defined charge distribution.

Experimental setup shown in Figure 2 was modified to evaluate NSAM response using NaCl

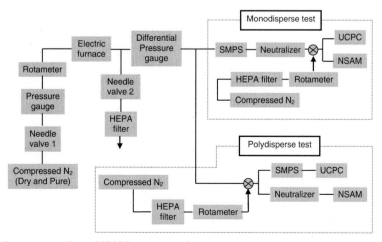

Figure 2. Experimental setup to evaluate NSAM response using monodisperse or polydisperse Ag particle agglomerates.

particles in a size range from 17 to 100 nm. A constant output atomizer (Model 3075, TSI, Inc.) was used to generate NaCl particles, and pure and dry air compressed at 37 psi was passed through the atomizer. In order to change particle size distributions, the concentration of NaCl solutions was varied from 0.0001 to 0.01 g/cc. The flow rate from the atomizer was about 3.0 lpm. Aerosol flow was passed through a diffusion dryer and then through a neutralizer to prevent particle loss due to highly charged status. In order to get higher monodispersity, primary particles larger than peak size of a given particle size distribution were subsequently selected with an electrostatic classifier (Model 3080, TSI, Inc.). In order to minimize the effect of humidity during the measurement, silica gel in the diffusion dryer was regenerated frequently. The humidity was kept as low as 30–50% during all experiments. Therefore, NaCl particles could not change in size during or after DMA classification (Tang et al., 1977). Splitting of the aerosol flow after the dilution was the same as in the case of Ag particle agglomerates.

The particle number concentration and electrometer current were simultaneously measured from UCPC and NSAM for all the experiments. The data integration time for all instruments was set to 2 min.

Normalized sensitivity and calibration factor of NSAM

For the NSAM, the input is particle number concentration (N) and the output (I) is electrometer current. Therefore, the size dependent sensitivity (S) of NSAM is given by

$$S(d_\mathrm{p}) = \frac{I(d_\mathrm{p})}{N(d_\mathrm{p})}, \qquad (1)$$

where d_p is the particle diameter. All sensitivity data are normalized with respect to the sensitivity for 100 nm particles. In other words, the sensitivity for 100 nm particles is used as a reference point. Normalized sensitivity (NS) is given by

$$NS(d_\mathrm{p}) = \frac{S(d_\mathrm{p})}{S(100\,\mathrm{nm})}. \qquad (2)$$

Calibration factor of NSAM is necessary to convert NSAM electrometer current signal into nanoparticle surface area deposited in human lung. Therefore, the calibration factor (CF) is given by

$$CF = \frac{\text{Surface area deposited in human lung}}{\text{Electrometer current}},$$

$$(3)$$

where CF has a unit of $\mu m^2/(cm^3 \ast pA)$.

Experimental results and discussion

Response function curves of NSAM for each d_p^2-weighted TB or A region are shown in Figures 3 and 4, respectively. Response function curves are plotted in terms of the normalized sensitivity. It is very useful to determine how well a response function curve of NSAM is matched with an ideally wanted response function curve obtained from a theoretical lung deposition efficiency curve for TB or A region. The deposition efficiencies in TB and A regions (η_{TB} and η_A) were obtained using the UK National Radiological Protection Board's (NRPB's) LUDEP Software (James et al., 2000), based on the recommendations of ICRP Publication 66 (ICRP, 1994). The lung deposition efficiency curves of spherical nanoparticles were derived for a reference worker (Fissan et al., 2006). For particles below 100 nm, particle density has no influence on the particle lung deposition efficiency curve because the dominant mechanism of particle lung deposition is only diffusion in the size range (Heyder et al., 1986).

As shown in Figures 3 and 4, our experimental data are compared with Fissan et al. (2006) results for EAD, which is essentially the same instrument as NSAM. For Ag particle agglomerates, the two experimental data sets show good agreement with each other for both the TB and A regions. Response function curves of NSAM are well matched with those of the ideally wanted lung deposition efficiency curves for both TB and A regions. There is a small difference between Ag particle agglomerates and NaCl particles in terms of normalized sensitivity.

Even though Ag particle agglomerates below 10 nm show larger deviation from the ideally wanted response function curve in Figures 3 and 4, it has no significant contribution to integrated surface area measurement deposited in human lung because of the d_p^2 dependency. The surface

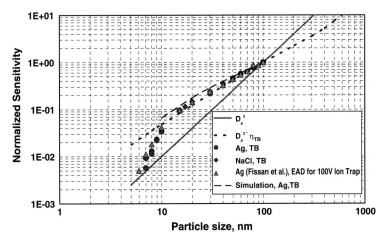

Figure 3. Comparison of response function curves of NSAM for TB region.

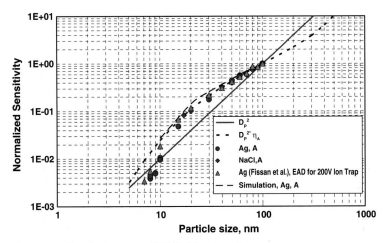

Figure 4. Comparison of response function curves of NSAM for A region.

area contribution of a 10 nm particle is only 1% of that of a 100 nm particle (Fissan et al., 2006).

Calibration curves of NSAM for TB is plotted in Figure 5. The calibration factor is obtained as the slope of the linear curve fitted to the data points on the plot for the particle surface area deposited in human lung vs. the NSAM electrometer current. Using the particle size measurement and the particle penetration efficiency, the particle surface area deposited in the human lung (DS) is calculated by

$$DS(d_p) = \pi d_p^2 \eta(d_p) \text{ for monodisperse particles},$$

$$DS = \Sigma \pi d_p^2 \eta(d_p) \text{ for polydisperse particles}, \quad (4)$$

where $\eta(d_p)$ is the particle lung deposition efficiency in TB or A region of human lung.

As shown in Figure 5, there exists a linear relation between the NSAM electrometer current signal and the particle surface area deposited in human lung. However, there may be a weak dependency of the calibration factor on the particle material. Calibration factors of NSAM for different parts of the inhalation system and different particle materials are summarized in Table 1. The differences of calibration factors between Ag particle agglomerates and NaCl particles are below 13% for both the TB and A regions. Several factors such as the particle morphology, the change of particle size, and particle material may have influence on this difference. However, it is believed that the first two factors have no influence on the difference of calibration factor in this experiment. Ku et al. (2005) showed that diffusion charging

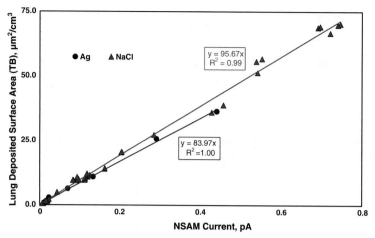

Figure 5. Calibration factor of NSAM for TB region.

Table 1. Summary of calibration factors ($\mu m^2/(cm^3\ pA)$) of NSAM

Region	Ag	NaCl
TB	83.97	95.67
A	385.24	422.68

responses of the LQ1-DC (Matter Engineering, Switzerland) and DC2000CE (EcoChem, USA) were proportional to the mobility diameter squared, regardless of the morphology for monodisperse Ag particle agglomerates below 100 nm. Based on the result, the NSAM response for Ag particle agglomerates is assumed to be very close to that for spherical monodisperse Ag particle. Also, NaCl particles could not be changed in size during or after DMA classification because humidity was kept lower than the deliquescent point above which NaCl particles only can grow (Tang et al., 1977). Our data suggest that the difference of 13% between Ag particle agglomerates and NaCl particle may be attributed to the weak dependence on materials. The dielectric constant of the particle material has a second order effect on particle charging rate (Liu & Pui, 1977).

The calibration result of NSAM using polydisperse Ag particles is in agreement with that using monodisperse Ag particles for a TB region (Figure 6). It demonstrates the huge linear range of the instrument. It also demonstrates that polydisperse NaCl–aerosol can be more easily used for calibration. Therefore, any monodisperse and polydisperse aerosols can be used for the calibration purpose of NSAM for both the TB and A regions, as long as the aerosols are limited in the size range where the response function is comparable with the needed response function.

Numerical simulation results of NSAM flow path and discussion

Using a commercial CFD s/w Fluent 6.2, numerical simulations of fluid flow and particle trajectories inside NSAM were performed to estimate response functions of NSAM under different ion trap voltages. As shown in Figure 1, the geometry inside NSAM is very complex and the charging process is extremely complicated, which involves turbulent mixing of airborne particles with unipolar ions produced from a corona needle. Therefore, the model of NSAM is simplified into flow path without the diffusion charger chamber. The simplified geometry model starts from the exit of the charger chamber and ends in front of the electrometer filter. It is assumed that the exit of the charger chamber exactly matches the inlet of the ion trap. Flow variables are uniform at the inlet boundary and the inlet velocity is 2.61 m/s. Navier–Stokes equations are solved using the implicit solver in Fluent 6.2. The flow is in the laminar flow regime. The second order upwind scheme is used for the velocity equations.

Derivation of the normalized sensitivity is necessary to compare experimental data with numerical simulation results. The current signal (I)

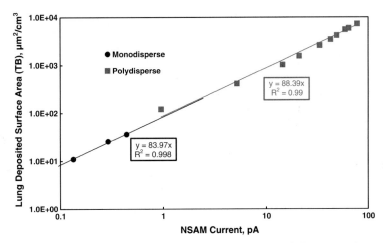

Figure 6. Comparison of calibration factors of NSAM for TB region using polydisperse and monodisperse Ag particle agglomerates.

measured by the electrometer filter inside the NSAM in the experiment is given by

$$I(d_{\mathrm{p}}) = N_{\mathrm{in}}(d_{\mathrm{p}})Q_{\mathrm{in}}n(d_{\mathrm{p}})e\eta_{\mathrm{tot}}(d_{\mathrm{p}}),\qquad(5)$$

where N_{in} is the particle number concentration at the inlet of NSAM, Q_{in} is the total flow rate (2.5 lpm) entering into the inlet of NSAM, $n(d_{\mathrm{p}})$ is the particle mean charge level, e is the charge on an electron, $\eta_{\mathrm{tot}}(d_{\mathrm{p}})$ is the total penetration efficiency of particle in NSAM.

The mean charge level per delivered particle from diffusion charger in EAD (Medved et al., 2000; Kaufman et al., 2002), which is proportional to the particle diameter for the size range of 10–100 nm, is used in the Fluent simulations. Equation (5) is substituted into Eq. (1) and the sensitivity (S) is given by

$$S(d_{\mathrm{p}}) = \frac{N_{\mathrm{in}}(d_{\mathrm{p}})Q_{\mathrm{in}}n(d_{\mathrm{p}})e\eta_{\mathrm{tot}}(d_{\mathrm{p}})}{N_{\mathrm{in}}(d_{\mathrm{p}})}$$
$$= Q_{\mathrm{in}}n(d_{\mathrm{p}})e\eta_{\mathrm{tot}}(d_{\mathrm{p}}).\qquad(6)$$

Eq. (6) is substituted into Eq. (2) and the normalized sensitivity (NS) is given by

$$\mathrm{NS}(d_{\mathrm{p}}) = \frac{Q_{\mathrm{in}}n(d_{\mathrm{p}})e\eta_{\mathrm{tot}}(d_{\mathrm{p}})}{Q_{\mathrm{in}}n(100\,\mathrm{nm})e\eta_{\mathrm{tot}}(100\,\mathrm{nm})}$$
$$= \frac{n(d_{\mathrm{p}})}{n(100\,\mathrm{nm})}\frac{\eta_{\mathrm{tot}}(d_{\mathrm{p}})}{\eta_{\mathrm{tot}}(100\,\mathrm{nm})}.\qquad(7)$$

The normalized sensitivity is expressed as the ratio of mean charge levels multiplied by that of

total penetration efficiencies. The total penetration efficiency is given by

$$\eta_{\mathrm{tot}}(d_{\mathrm{p}}) = \eta_{\mathrm{flowpath}}(d_{\mathrm{p}}) \times \eta_{\mathrm{charger\,chamber}}(d_{\mathrm{p}}),\quad(8)$$

where $\eta_{\mathrm{flowpath}}(d_{\mathrm{p}})$ and $\eta_{\mathrm{charger\,chamber}}(d_{\mathrm{p}})$ are the penetration efficiencies of particles through the flow path and the charger chamber, respectively.

Equation 8 is substituted into Eq. (7) and the normalized sensitivity is given by

$$\mathrm{NS}(d_{\mathrm{p}}) = \frac{n(d_{\mathrm{p}})}{n(100\,\mathrm{nm})}\frac{\eta_{\mathrm{flowpath}}(d_{\mathrm{p}})}{\eta_{\mathrm{flowpath}}(100\,\mathrm{nm})}$$
$$\times \frac{\eta_{\mathrm{charger\,chamber}}(d_{\mathrm{p}})}{\eta_{\mathrm{charger\,chamber}}(100\,\mathrm{nm})}.\qquad(9)$$

If the ratio of the penetration efficiencies for the charger chamber in the Eq. (9) is assumed to be equal to 1, the normalized sensitivity (NS) is simplified as

$$\mathrm{NS}(d_{\mathrm{p}}) = \frac{n(d_{\mathrm{p}})}{n(100\,\mathrm{nm})}\frac{\eta_{\mathrm{flowpath}}(d_{\mathrm{p}})}{\eta_{\mathrm{flowpath}}(100\,\mathrm{nm})}.\qquad(10)$$

The flow path penetration efficiency in Eq. (10) was obtained through particle trajectory calculations using the discrete phase model in Fluent. The effect of Brownian motion was included in the simulations. Ag particles used in Fluent simulations are in the range of 10–100 nm. In order to investigate the electrophoresis effect, a user-defined subroutine was run during the Fluent calculation. It adds the electric force to the existing body force term of the discrete phase model in Fluent and thus enables the application of the electric field in the ion trap part. The mean charge level of particles injected from the inlet boundary is set to be proportional to the particle size in the user-defined subroutine. In each particle trajectory calculation, 1000 spherical Ag particles with the same size were released from the inlet boundary and each particle has the mean charge level proportional to its size. Material properties of Ag include a density of 10.5 kg/m^3, a thermal conductivity of 429 W/m K, and a specific heat of 235 J/kg K. The penetration efficiency for the flow path was calculated by dividing the number of particles arriving at the end point by the number of injected particles. For each particle size, the procedure of particle trajectory calculation was repeated over 20 times to improve the statistical accuracy. The normalized sensitivity was obtained by multiplying the ratio of the penetration efficiency for the flow path determined from numerical simulations by that of mean charge level as shown in Eq. (10).

A comparison of the response function curves obtained by Fluent simulation and experimental measurement can be found in Figures 3 and 4. Fluent simulation results are relatively well matched with experimental results in the size range between 10 and 100 nm. Because particles below 10 nm do not contribute significantly to the particle surface area deposited in human lung, the particles were not modeled. Normalized sensitivity estimated by Fluent simulation is larger than that obtained by experimental measurement in both cases of TB and A regions. The deviation between the model and the experimental results points to the importance of the charger chamber for the whole system. With the next version of the model, the effects within the charger chamber will be modeled.

Conclusions and discussion

We have experimentally investigated the response functions and calibration factors of a modified EAD, the TSI Model 3550 Nanoparticle Surface Area Monitor (NSAM), for the measurement of nanoparticle surface area deposited in the human lung. Also, we have analytically derived normalized sensitivity of the NSAM to compare experimental data with numerical simulation results. The comparison indicates that numerical simulation is a valuable method to predict the NSAM results.

NSAM shows a linear relation between the particle surface area deposited in human lung and the electrometer current for both TB and A regions of a reference worker. The results look promising in that NSAM gives a real-time quantitative measurement of the particle surface area deposited the human lung and thus can be used to correlate with epidemiological studies for nanoparticles below 100 nm. From the experimental data, there is a weak dependency of response function curves of NSAM on particle materials. Also, the NSAM calibration factors for both the TB and A regions show a weak dependency on particle materials. The differences between Ag particle agglomerates and NaCl particles are below 13%. For calibration, monodisperse nanoparticles in the size range where the response function is in the desirable range can be used. Polydisperse nanoparticles can be used more easily for the calibration purpose.

Fluent simulations results matched reasonably well with experimental results in terms of the response function. It demonstrates that Fluent simulation can be used to optimize the NSAM response qualitatively. However, in order to get more accurate results quantitatively from numerical simulation, the charger chamber penetration efficiency also needs to be considered in the numerical simulations. Furthermore, it is suggested that modification of ion trap configurations can be considered to improve the match between the response function of NSAM and the ideally wanted response function curve.

Acknowledgements

The authors wish to acknowledge the University of Minnesota Supercomputer Institute for providing the computation time for this project.

References

Baltensperger U., H.W. Gäggeler & D.T. Jost, 1988. The epiphaniometer, a new device for continuous aerosol monitoring. J. Aerosol Sci. 19(7), 931–934.

Brunauer S., P.H. Emmett & E. Teller, 1938. Adsorption of gases in multimolecular layers. J. Am Chem. Soc. 60, 309–319.

Donaldson K., X.Y. Li & W. MacNee, 1998. Ultrafine (nanometer) particle mediated lung injury. J. Aerosol Sci. 29(5–6), 553–560.

Fissan H. & T. Kuhlbusch, 2005. Strategies and instrumentation for nanoparticle exposure control in air at workplaces. 2nd International symposium on nanotechnology and occupational health, Minneapolis, USA, October 3–6, 2005.

Fissan H., A. Trampe, S. Neunman, D.Y.H Pui & W.G. Shin, 2006. Rationale and principle of an instrument measuring lung deposition area. J. Nanoparticle Research (this issue).

Han H.S., D.R. Chen, B.E. Anderson & D.Y.H. Pui, 2000. A nanometer aerosol size analyzer (nASA) for rapid measurement of high concentration size distributions. J. Nanoparticle Res. 2, 43–52.

Heyder J., J. Gebhart, G. Rudolf, C.F. Schillerd & W. Stahlhofen, 1986. Deposition of particles in the human respiratory tract in the size range 0.005–15µm. J. Aerosol Sci. 17, 811–825.

ICRP., 1994. International Commission on Radiological Protection Publication 66 Human Respiratory Tract Model for Radiological Protection. Oxford, Pergamon: Elsevier Science Ltd.

James A.C., M.R. Bailey & M-D. Dorrian, 2000. LUDEP Software, Version 2.07: Program for implementing ICRP-66 Respiratory tract model. RPB, Chilton, Didcot, OXON. OX11 ORQ UK.

Jung H.J. & D.B. Kittelson, 2005. Characterization of aerosol surface instruments in transition regime. Aerosol Sci. Tech. 39(9), 902–911.

Kaufman S.L., A. Medved, A. Pöcher, N. Hill, R. Caldow & F.R. Quant, 2002. An electrical aerosol detector based on the corona-jet charger. AAAR conference (poster).

Ku B.K. & A.D. Maynard, 2005. Generation and investigation of airborne Ag nanoparticles with specific size and morphology by homogeneous nucleation, coagulation and sintering. J Aerosol Sci. 36(9), 1108–1124.

Liu B.Y.H. & D.Y.H. Pui, 1977. On unipolar diffusion charging of aerosols in the continuum regime. J. Colloid Interface Sci. 58, 142–149.

Maynard A.D. & E.D. Kuempel, 2005. Airborne nanostructured particles and occupational health. J. Nanoparticle Res. 7(6), 587–614.

Maynard A.D., 2003. Estimating aerosol surface area from number and mass concentration measurements. Ann. Occup. Hygiene 47, 123–144.

Medved A., F. Dorman, S.L. Kaufman & A. Pöcher, 2000. A new corona-based charger for aerosol particles. J. Aerosol Sci. 31(S.1), 616–617.

Oberdorster G., R.M. Gelein, J. Ferin & B. Weiss, 1995. Association of particulate air pollution and acute mortality: involvement of ultrafine particles. Inhal. Toxicol. 7, 111–124.

Oberdorster G., 1996. Significance of particle parameters in the evaluation of exposure-dose-response relationships of inhaled particles. Particulate Sci. Technol. 14(2), 135–151.

Oberdörster G., E. Oberdörster & J. Oberdörster, 2005. Invited review: nanotechnology: an emerging discipline evolving from studies of ultrafine particles. Environ. Health Perspect. 113(7), 823–839.

Shi J.P., R.M. Harrison & D. Evans, 2001. Comparison of ambient particle surface area measurement by epiphaniometer and SMPS/APS. Atmos. Environ. 35(35), 6193–6200.

Tang I.N., H.R. Munkelwitz & J.G. Davis, 1977. Aerosol growth studies—II. Preparation and growth measurements of monodisperse salt aerosols. J. Aerosol Sci. 8(3), 149–159.

Wilson W.E., H.-S. Han, J. Stanek, J. Turner & D.Y.H. Pui, 2003. The Fuchs surface area measured by charge acceptance of atmospheric particles may be a useful indicator of the quantity of particle surface area deposited in the lung. Abstracts of the European aerosol conference. S421-S422. Madrid, Spain.

Wilson W.E., H.-S. Han, J. Stanek, J. Turner, D.-R. Chen & D.Y.H. Pui, 2004. Use of electrical aerosol detector as an indicator for the total particle surface area deposited in the lung. Symp. On air quality measurement methods and technology sponsored by air and waste management association. Research triangle park, NC. Paper #37.

Woo K.-S., D.-R. Chen, D.Y.H. Pui & W.E. Wilson, 2001. Use of continuous measurements of integral aerosol parameters to estimate particle surface area. Aerosol Sci. Tech. 34, 57–65.

Journal of Nanoparticle Research (2007) 9:71–83
DOI 10.1007/s11051-006-9152-z

Special focus: Nanoparticles and Occupational Health

© Springer 2006

An axial flow cyclone to remove nanoparticles at low pressure conditions

Sheng-Chieh Chen and Chuen-Jinn Tsai
Institute of Environmental Engineering, National Chiao Tung University, Hsin Chu, 300, Taiwan;
(E-mail: cjtsai@mail.nctu.edu.tw)

Received 25 July 2006; accepted in revised form 28 July 2006

Key words: axial flow cyclone, nanoparticle control, particle loading effect, particle control equipment, occupational health

Abstract

In this study, the axial flow cyclone used in Tsai et al. (2004) was further tested for the collection efficiency of both solid (NaCl) and liquid (OA, oleic acid) nanoparticles. The results showed that the smallest cutoff aerodynamic diameters achieved for OA and NaCl nanoparticles were 21.7 nm (cyclone inlet pressure: 4.3 Torr, flow rate: 0.351 slpm) and 21.2 nm (5.4 Torr, 0.454 slpm), respectively. The collection efficiencies for NaCl and OA particles were close to each other for the aerodynamic diameter ranging from 25 to 180 nm indicating there was almost no solid particle bounce in the cyclone. The 3-D numerical simulation was conducted to calculate the flow field in the cyclone and the flow was found to be nearly paraboloid. Numerical simulation of the particle collection efficiency based on the paraboloid flow assumption showed that the collection efficiency was in good agreement with the experimental data with less than 15% of error. A semi-empirical equation for predicting the cutoff aerodynamic diameter at different inlet pressures and flow rates was also obtained. The semi-empirical equation is able to predict the cutoff aerodynamic diameter accurately within 9% of error. From the empirical cutoff aerodynamic diameter, a semi-empirical square root of the cutoff Stokes number, $\sqrt{St^*_{50}}$, was calculated and found to be a constant value of 0.241. This value is useful to the design of the cyclone operating in vacuum to remove nanoparticles.

Introduction

Cyclones are normally used to remove particles larger than 5–10 μm in aerodynamic diameter. To reduce the cutoff diameter, the cyclone diameter must be reduced or the flow rate must be increased. For example, Zhu and Lee (1999) tested a small tangential cyclone (cyclone diameter, $D = 3.05$ cm) and found the cutoff size to be 0.3 μm when it was operated at 110 slpm (standard L/min). Making use of large slip correction factor of nanoparticles at reduced pressure, Tsai et al. (2004) designed and tested an axial flow cyclone to remove nanoparticles smaller than 100 nm at low pressure conditions. The smallest cutoff

aerodynamic diameter in the study is 43.3 nm when the inlet pressure of the cyclone is 6 Torr at the flow rate of 0.455 slpm. However, the experimental results of the collection efficiency are for liquid particles only. The solid particles collection efficiency of the cyclone has not been investigated.

Maynard (2000) is the first to study the particle penetration of the axial flow cyclone in ambient pressure theoretically. He proposed an implicit equation of the particle penetration based on the assumption that particle collection mainly occurs in the vane and body sections. The equation predicts the cutoff aerodynamic very well. Tsai et al. (2004) derived a theoretical equation for the particle collection efficiency based on the air volumetric flow rate, the geometry of the cyclone, the properties of carrying gas and the pressure of the cyclone. The cutoff aerodynamic diameter D_{pa50} for the cyclone with one vane making three turns was derived as

$$D_{pa50} = \frac{0.106\mu(B-w)\left(r_{max}^2 - r_{min}^2\right)^2}{\rho_{p0}\lambda_0 r_{min}^2 Q_0} \times \left(\frac{P_{cyc}}{P_{760}}\right)^2. \tag{1}$$

In the above equation, μ is the fluid dynamic viscosity (N s/m^2), B is the pitch of vanes (m), w is the vane thickness (m), r_{max} is the inner radius of the cyclone (m), r_{min} is the radius of the vane spindle (m), ρ_{p0} is unit density (1000 kg/m^3), λ_0 is the mean free path of air molecules at standard condition (m), Q_0 is standard volumetric flow rate (m^3/sec), P_{cyc} is the average pressure of cyclone inlet and outlet (Torr), P_{760} is 760 Torr. The equation agrees well with the published experimental data on cutoff diameter (Liu & Rubow, 1984; Weiss et al., 1987; Vaughan, 1988) in ambient conditions. But at low pressure conditions (several Torrs), the equation predicts the cutoff diameter much smaller than the experimental data. The theoretical cutoff diameter has to be adjusted to fit the experimental data. The reason why there is such as discrepancy has yet to be found.

Hsu et al. (2005) derived a model considering both centrifugal and diffusional forces for nanoparticle removal in vacuum using the axial flow cyclone of Tsai et al. (2004). They found that diffusional mechanism was important when particles were smaller than 40 nm.

For impactors, it is well-known that the collection of liquid particles is better than that of solid particles due to solid particles bounce or reentrainment from the impactor substrates (Biswas & Flagan, 1988; Tsai & Cheng, 1995; Tsai & Lin, 2000). Zhu and Lee (1999) studied the differences in the collection efficiency of a tangential flow cyclone for solid and liquid particles. Liquid dioctyl-phthalate (DOP) and solid polystyrene latex (PSL) particles in the size range of 1.0 and 3.6 µm were found to have similar collection efficiencies even at a high flow rate of 80 slpm. However, the effect of solid particle bounce on the collection efficiency of the axial flow cyclone operating in low pressure conditions remains to be investigated.

The effect of deposited solid particles on the cyclone wall of the tangential flow cyclone has been investigated in Blachman and Lippmann (1974) and Tsai et al. (1999). The particle collection efficiency was found to increase with increasing particle mass deposited in the cyclone (Blachman & Lippmann, 1974). Such increase is mainly due to the accumulation of dust on the cyclone wall opposite to the inlet which gradually reduces the effective diameter of the cyclone. But when the amount of deposited particles is heavy enough, the aggregated particles will be detached and then the collection efficiency will reduce again (Blachman & Lippmann, 1974). The effect of different amounts of deposited particle mass on the collection efficiency was not studied in Blachman and Lippmann (1974). The solid particle loading effect on the collection efficiency for a 10 mm nylon cyclone and a new 18 mm aluminum cyclone was studied by Tsai et al. (1999). They found the cutoff aerodynamic diameter of both cyclones decreased with increasing deposited particle mass. But the 18 mm cyclone appeared to have less deposited particle mass effect on the collection efficiency due to its larger inner diameter. The cyclones tested in Blachman and Lippmann (1974) and Tsai et al. (1999) were tangential flow cyclones. There have been no studies on the solid particle loading effect on the collection efficiency for axial flow cyclones.

The flow field in the cyclone is complicated. Several researchers have studied the flow fields numerically for tangential flow cyclones and examined the influence of different geometries and operating conditions on the collection efficiency of the cyclones (Boysan et al., 1983; Hoekstra et al., 1999; Schmidt & Thiele, 2002; Harwood & Slack, 2002; Schmidt et al., 2003; Xiang & Lee, 2004). However, there have been no numerical studies on the flow field and particle collection efficiency of axial flow cyclones, in particular at low pressure conditions.

In this study, NaCl and OA monodisperse nanoparticles in diameter from 12 to 100 nm were generated to test the axial flow cyclone of Tsai et al. (2004). The test conditions are: inlet pressure of 4.3 Torr at 0.351 slpm, 5.4 or 6.0 Torr at 0.455 slpm, and 6.8 or 7.0 Torr at 0.566 slpm.

The 3-D numerical simulation was also conducted in this study to calculate the flow and pressure fields of the cyclone in order to obtain a more accurate prediction of the collection efficiency and cutoff size of the cyclone.

Finally, a new empirical equation for predicting the cutoff aerodynamic diameter of the cyclone was also proposed based on the better knowledge of the 3-D flow and pressure fields obtained in this study.

Experimental

The experimental system is shown in Figure 1. Monodisperse OA ($\rho_p = $ 894 kg/m^3) and NaCl ($\rho_p = $ 2200 kg/m^3) particles in diameter between 12 and 100 nm were generated by the atomization and electrostatic classification technique. Polydisperse particles were first generated by atomizing (Atomizer, TSI Model 3076) 0.05 or 0.1% (v/v) OA and NaCl solution. Then the aerosol flow was dried by a silica gel drier. The dried aerosol stream was passed through a furnace (Lindberg/Blue Model CC58114C-1) and mixed with clean air to produce sufficiently small particles ($<$100 nm) after the furnace. The temperature of the furnace

was fixed at 650 and 1150 K for OA and NaCl particles, respectively. Fine polydisperse particles were generated by mixing the vapor with dry compressed air. Monodisperse, singly charged particles were generated by classifying the polydisperse particles by a nano-DMA (TSI Model 3085).

The SMPS (Condensation Particle Counter, TSI Model 3022 and Electrostatic Classifier, TSI Model 3071) was used to monitor the concentrations of particles in the monodisperse particle stream from the nano-DMA. The concentrations were used to correct for the multiple charge effect on the collection efficiency. For the detailed procedure, refer to Tsai et al. (2004).

An aerosol electrometer (TSI Model 3068) was used to measure the electric current of the upstream and downstream aerosol concentrations of the cyclone. A homemade Faraday cage with a larger inlet and outlet than the TSI Model 3068 electrometer was used to reduce the pressure drop through it. A critical orifice (O'Keefe Controls

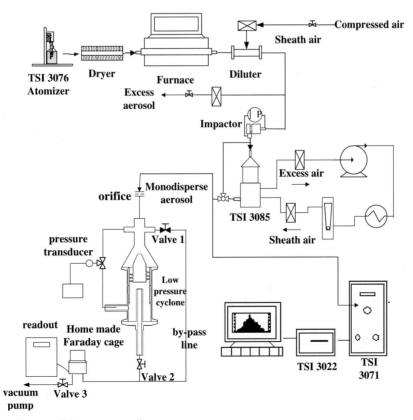

Figure 1. Experimental setup of the present study.

Co., E-8, 0.351 slpm or E-9, 0.455 slpm or E-10, 0.566 slpm) was installed at the cyclone inlet to achieve the low-pressure condition. A powerful vacuum pump (DUO 65, Pfeifeer, Germany, nominal pumping speed: 70 m³/h) was used to achieve the desired low pressure condition. The inlet pressures at the cyclone inlet in this study are 4.3, 6.0 5.4, 6.8 and 7 Torr.

The bypass line was used to determine the particle concentration at the cyclone inlet which can be controlled by an on-off valve (Valve 1) as shown in Figure 1. When Valve 1 is open and Valve 2 is closed, the aerosol flow will pass through the bypass line and the inlet aerosol concentration can be measured. On the other hand, when Valve 1 is closed and Valve 2 is open, the aerosol flow will pass through the cyclone and the particle concentration at the cyclone outlet can be obtained. By adjusting the angle valve (Valve 3) at the downstream of the Faraday cage, the pressure at the cyclone inlet can be controlled.

The loading effect test was conducted by introducing polydisperse particles continuously into the cyclone over a period of time. The particle collection efficiency was tested after loading polydisperse particles (total number conc.: 8.26×10^6–$1.2 \times 10^7 \#/cm^3$, NMD: 69.5–82 nm, σ_g: 1.53–1.58) for 1-h (loaded mass: 0.33 mg), 3-h (1.24 mg), and 5-h (1.73 mg), respectively.

The spindle and vane of the cyclone are shown in Figure 2. The radius of the spindle and the inner radius of the cyclone are 10 and 15 mm, respectively. The width and the height of the vane section are 5 and 4 mm, respectively.

Numerical

Flow field

In order to obtain accurate pressure distribution and gas velocity fields in the cyclone, 3-D numerical simulation was conducted in the present study. The governing equations are Navier–Stokes and the continuity equations. Since the maximum Kundsen number in the present study for the pressure of 1.46 Torr (P_{in} = 4.31 Torr) is around 0.01, the flow was considered as continuum. The maximum Kundsen number, Kn,max is calculated as

$$Kn, \max = \frac{2\lambda}{D_o}, \qquad (2)$$

where D_o is the inner diameter of the outlet tube (m); λ is the mean free path of the air molecular (m). Moreover, Steady-state and compressible laminar flow was assumed in this study. The Navier–Stokes and the continuity equations were solved by using the STAR-CD 3.15 code (CD-adapco Japan Co., LTD) based on the finite volume discretization method. The pressure-velocity linkage was solved by the SIMPLE (semi-implicit method for pressure linked equation) algorithm (Pantankar, 1980) and the differencing schemes for the space discretization method were the UD (upwind differencing) and CD (central differencing) schemes for the flow velocity and density, respectively.

Multi-block, hybrid (both Hexa. and Tetra.) cells were generated by an automatic mesh generation tool, Pro-Modeler 2003 (CD-adapco Japan Co., LTD). The total number of cells used was about 1,000,000. The average cell length was around 0.5 mm and the smallest length of 0.1 mm was assigned near the surface of the vane.

The convergence criterion of the flow field calculation was set to be 0.01 % for the summation of the residuals. The total number of iterations was about 300 and the time required to reach convergence was about 600 min. Non-slip condition was applied on the walls. A constant mass flow rate (0.351, 0.455, or 0.566 slpm) was set on the inlet boundary assuming uniform velocity profile. On the outlet boundary, a fixed pressure was assigned based to the experimental data.

Particle collection efficiency

As a particle enters the cyclone, it experiences the centrifugal force and migrates toward the wall. The Stokes law was adopted to calculate the particle drag force since the Rep (particle Reynolds number) was much small than 0.1 in this study. Based on the assumption that particle drag force is equal to the centrifugal force, the particle radial migration velocity, V_r, is calculated as

$$V_r = \frac{\tau V_t^2}{r}, \qquad (3)$$

where V_t is the tangential flow velocity, r is the radial position of the particle and τ is the

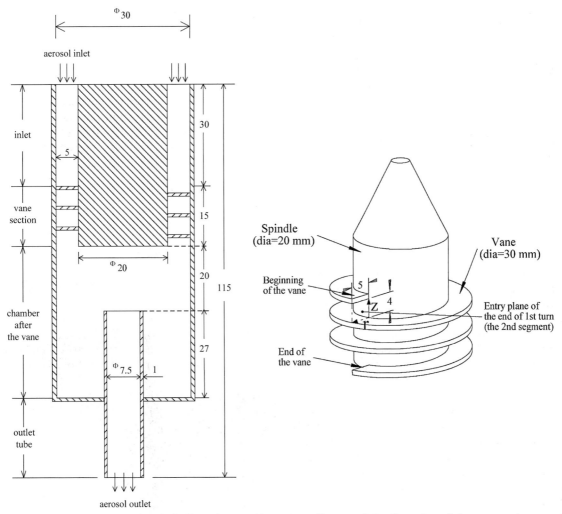

Figure 2. Schematic diagram of the spindle and vane. (the *r–z* coordinate and the dimension of the vane section are also indicated.)

particle relaxation time, which can be written as

$$\tau = \frac{\rho_{p0} D_{pa}^2 C(D_{pa})}{18\mu}, \tag{4}$$

where D_{pa} and $C(D_{pa})$ are the particle aerodynamic diameter and the slip correction factor, respectively. Both τ and V_t are the function of pressure in the cyclone.

The differential radial migration distance of the particle, dr, can be calculated as

$$\mathrm{d}r = V_r dt = \frac{\tau V_t^2}{r} \frac{r\mathrm{d}\theta}{V_t} = \tau V_t \mathrm{d}\theta, \tag{5}$$

The paraboloid flow was assumed to calculate the particle migration distance. Since the flow in the cyclone was found to spin for slightly greater than 2 turns starting slightly ahead of the end of the first turn and ending slightly beyond the end of the third turn, only the particle migration distance during 2 turns was calculated. To simplify the calculation, the total radial migration distance of a particle, Δr, was calculated based on Eq. (5) as the sum of the migration distance from ten segments of the vane section, each segment corresponded to 1/4 turn of the vane. The first segment or the first 1/4 turn was added before the end of the first turn while another 1/4 turn was added after the end of

76

the third turn. The second and third turn each constituted four segments in the calculation. All together there were 10 segments.

As will be shown later, the tangential flow develops very fast and becomes nearly fully developed near the end of the first turn of the vane. The fully developed profile is paraboloid which can be written as

$$V_{t,n^{th}}(r,z) = 2(\overline{V}_{t,n^{th}} - 1)\left[1 - \left(\frac{2z}{4}\right)^2\right] \times$$

$$\left[1 - \left(\frac{2r}{5}\right)^2\right] + 2, \quad \text{m/s} \qquad (6)$$

where the coordinates r and z are illustrated in Figure 2. $V_{t,n^{th}}(r,z)$ and $\overline{V}_{t,n^{th}}$ are the tangential velocity of the entry plane of the nth segment at position (r,z) and the average tangential velocity of the nth segment, respectively. The constant 2 m/s at the right-hand side of Eq. (6) represents the tangential velocity near the wall, which is obtained from the numerical simulation shown later. If the total migration distance of a particle of aerodynamic diameter D_{pa} plus the initial radial position is greater than 5 mm (or $r_{max}-r_{min}$, the width of the vane section) then the particle hits the wall and is collected. Assuming different initial radial positions of a particle at the entry plane of the first segment, the critical curve which delineates the collection and non-collection regions of the particle can be found. As the collection area is obtained, then the collection efficiency can be calculated by the following equation as

$$\eta_{dpa} = \frac{A \times \overline{V}'_{t,1}}{5 \times 4 \times \overline{V}_{t,1}}, \qquad (7)$$

where A is the collection area (mm²); $\overline{V}'_{t,1}$ is the average tangential velocity of the collection area (m/s); 5 (mm) and 4(mm) are the width and gap of the vane section, respectively (mm); $\overline{V}_{t,1}$ is the average tangential velocity at the entry plane of the 1st segment of the vane section.

Since the pressure drop of the cyclone occurs mainly in the vane section, $\overline{V}_{th,n^{th}}$ is calculated based on the pressure at the nth section following the ideal gas law and mass conservation principle. For comparison purpose, if the tangential flow field is assumed to be plug flow, the total radial migration distance Δr can be calculated as (referring to Eq. (5))

$$\Delta r = \sum_{n^{th}=1}^{10} \frac{\pi}{2} \overline{\tau}_{n^{th}} \overline{V}_{t,n^{th}}, \qquad (8)$$

where $V_{t,n^{th}}$ is the average tangential velocity (m/s), $\overline{\tau}_{n^{th}}$ is the average relaxation time of the particle. Both $V_{t,n^{th}}$ and $\overline{\tau}_{n^{th}}$ depend on the average pressure at the nth segment. The collection efficiency η of the particle can be calculated as

$$\eta = \frac{\Delta r}{r_{max} - r_{min}}. \qquad (9)$$

Results and discussion

Comparison of collection efficiency for liquid and solid particles

The collection efficiencies of solid NaCl and liquid OA particles for the inlet pressure of 6 and 5.4 Torr and the sampling flow rate of 0.455 lpm are compared in Figure 3. For OA particles, the present experimental data are in good agreement with Tsai et al. (2004). The collection efficiencies are seen to be greatly improved for both OA and

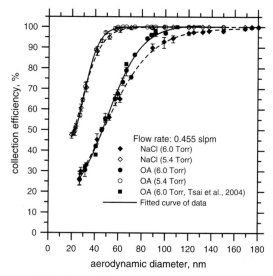

Figure 3. Collection efficiency *versus* particle aerodynamic diameter for solid NaCl and liquid OA particles at different inlet pressures.

NaCl particles when the pressures at the cyclone inlet are reduced from 6 to 5.4 Torr. The cutoff aerodynamic diameters of OA and NaCl particles are reduced from 49.8 and 47.1 to 23.1 and 21.2 nm, respectively as the pressure at the cyclone inlet is decreased from 6 to 5.4 Torr, respectively. In addition, Figure 3 indicates that the collection efficiencies of liquid and solid particles are close to each other for both operation pressures except in the size range from 60 to 120 nm at 6 Torr where the collection efficiency of OA is slightly better (within 10%) than NaCl. This is to say that the effect of solid particle bounce on the collection efficiency is not very obvious. Furthermore, it also found the diffusion effect on collection efficiency for particles less than (40 nm) is not as significant as claimed by Hsu et al. (2005).

Solid particle loading effect on collection efficiency

The effect of polydisperse particle loading on the collection efficiency is shown in Figure 4 for the inlet pressure of 6 and 5.4 Torr, respectively. For both operating pressures, the collection efficiency after 1-h loading (loaded mass: 0.33 mg) is 5–10 % higher than that of a clean cyclone (or zero particle loading). However, the collection efficiency after 3-h (1.24 mg), and 5-h (1.73 mg) loading is not too much different from that of 1-h loading.

Tsai et al. (1999) used two monodisperse particles, 3.76 and 6.7 μm in aerodynamic diameter, to examine the solid loading effect on the collection efficiency of a 10 mm nylon cyclone. Both loaded particle masses in the cyclone were 0.06 mg. Their experimental results indicated that the particle penetration decreased (or collection efficiency increased) for both loading conditions compared to that of the clean condition. The present results also show that the collection efficiency increases after particles are loaded in the cyclone. However, the increase in the collection efficiency is larger in Tsai et al. (1999) (10–30 % increases, small 10 mm tangential flow cyclone) than the present study (5–10 % increases, axial flow cyclone). This is due to the accumulation of deposited particles on the cyclone wall opposite to the inlet for the small tangential flow cyclone, which has a larger influence on the collection efficiency. In comparison, the present axial flow cyclone has a larger cyclone diameter and a more uniform particle deposit on the cyclone wall. As a result, the solid particle loading effect on the collection efficiency is not as significant as that of the small tangential flow cyclone.

The collection efficiency of liquid OA particles at 5 different inlet pressures is shown in Figure 5. The flow rate and the Reynolds number range from 0.351 to 0.566 slpm and 4.9 to 8.0, respectively. The Reynolds number, *Re*, is defined as

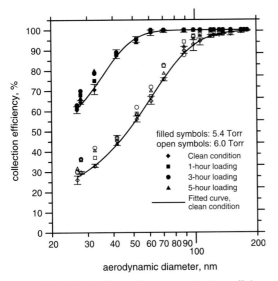

Figure 4. Particle loading effect on collection efficiency, solid NaCl particles.

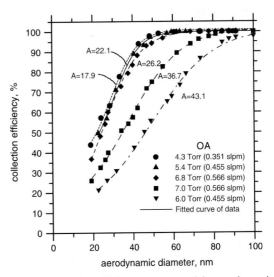

Figure 5. Collection efficiency *versus* particle aerodynamic diameter for OA particles at different inlet pressures and flow rates. (parameter $A = P_{in} \times P_{out}/Q_0$).

$$Re = \frac{\rho(r_{max} - r_{min})\overline{V}_a}{\mu}, \qquad (10)$$

where $\rho, \overline{V}_a$ are the gas density (kg/m^3) and the average axial velocity (m/s), respectively. $\rho, \overline{V}_a$, and μ were evaluated at the entry of the vane section. The figure shows that the parameter A, which is defined as $A = P_{in} \times P_{out}/Q_0$ (P_{in}: pressure at the cyclone inlet; P_{out}: pressure at the vane outlet), influences the collection efficiency and cutoff diameter. The larger A is the smaller collection efficiency the cyclone becomes. The cutoff size is 21.7, 23.1, 25.6 nm for the inlet pressure of 4.3 Torr at 0.351 slpm ($A = 17.93$), 5.4 Torr at 0.455 slpm ($A = 22.08$), and 6.8 Torr at 0.566 slpm ($A = 26.19$) (also shown in Table 1). The cutoff size becomes much larger for the inlet pressure of 7.0 Torr at 0.566 slpm ($A = 36.73$), 6.0 Torr at 0.455 slpm ($A = 43.12$) as the parameter A becomes much larger for these two cases (also shown in Table 1). The reasons why the cutoff size is affected by A will be explained later when the empirical equation for cutoff size is derived.

Numerical results for flow field and particle collection efficiency

The simulated results for the pressure distribution and the maximum tangential velocity in the vane section for $P_{in} = 5.4$ Torr (0.455 slpm) are shown in Figure 6. It can be seen that the tangential velocity remains small in the beginning of the first turn of the vane. However, it increases sharply from the entry plane of the first segment (or the beginning of 3/4 turn). Then it increases exponentially until the end of 3 turns (or the ninth segment) and reduces sharply in the tenth segment.

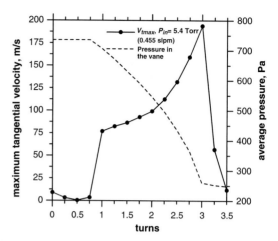

Figure 6. Maximum tangential velocity and pressure distribution in the vane, numerical results.

The tangential velocity peaks at 187 m/s at the end of 3 turns and is reduced to 60 m/s sharply at the end of 10th segment.

The pressure distribution shown in Figure 6 corresponds well with the tangential velocity distribution. The flow is accelerated in the tangential direction as the pressure is decreased in the vane. The pressure remains at about 735 Pa (5.5 Torr, within 2% of the experimental data) in the first turn and then drops monotonically in the second and third turns. Finally, the pressure remains nearly constant at 260 Pa (1.95 Torr) after the end of the third turn. That is, the pressure drop occurs almost entirely in the second and third turns. The figure shows that the flow does not make three full turns as the vane does. Rather, it makes slightly more than two turns only. This is one of the main reasons why the differences are large between the experimental cutoff aerodynamic diameters and the theoretical values in Tsai et al. (2004) in which

Table 1. Cutoff aerodynamic diameter at different operating conditions, liquid oleic acid (OA) particles

Q_0 (slpm)	0.351 Re = 4.9	0.455 Re = 6.4	0.566 Re = 8.0	0.566 Re = 8.0	0.455 Re = 6.4
p_{in} (Torr)	4.31	5.43	6.77	7.00	6.00
p_{out} (Torr)	1.46	1.85	2.19	2.97	3.27
$A = P_{in} \times P_{out}/Q_0$	17.93	22.08	26.19	36.73	43.12
Exp. D_{pa50}(nm)	21.69	23.14	25.58	34.71	46.25
Num$_{parab}$ D_{pa50}(nm)	20.4	23.3	25.21	31.2	39.8
Num$_{plug}$ D_{pa50}(nm)	18.1	20.2	23.6	30.6	34.3
Theo. D_{pa50} (nm), Eq. (8)	13.2	16.5	19.3	27.1	31.8
K (Exp./Theo.)	1.64	1.4	1.33	1.28	1.45

the tangential flow was assumed to make three full turns and the tangential velocity was assumed to be plug flow.

Figure 7 a, b show the tangential velocity profile at the vertical cut planes at the end of 2 and 3 turns, respectively. The tangential velocity peaks near the center of the plane in both figures and the value is about two times the average tangential velocity, which was given in Tsai et al. (2004) as

$$\overline{V}_t = \frac{2r_{\min}Q}{(r_{\max}^2 - r_{\min}^2)(B - w)}, \qquad (11)$$

where Q is the volumetric flow rate and can be calculated as

$$Q = Q_0 \frac{P_{760}}{\sqrt{P_{\text{in}}P_{\text{out}}}}. \qquad (12)$$

It is also found that the velocity near the wall is not zero but about 2 m/s. Similar results can be found in other cut planes and at other different operating conditions ($P_{\text{in}} =$ 4.3, 6, 6.8 and 7 Torr). From the simulated tangential velocity shown in Figure 7a and b, the velocity profile is found to be nearly paraboloid, and the variation of velocity with r and z positions can be calculated by Eq. (6).

Figure 8a shows the critical collection curves for a 23 nm particle passing through the full three

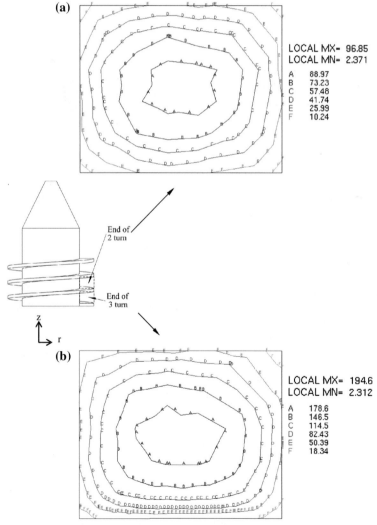

Figure 7. Tangential velocity profile at the cut plane of the vane section, the end of (a) two turns and (b) three turns.

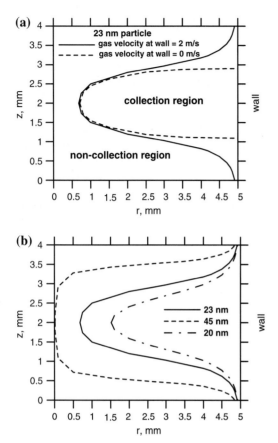

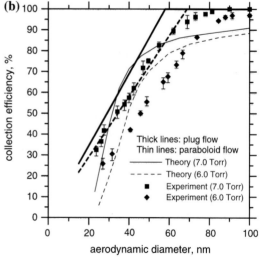

Figure 8. Critical curves at the entry plane of the first segment for particle collection. (a) With or without a constant tangential velocity of 2 m/s at the wall for a 23 nm particle. (b) With a constant tangential velocity of 2 m/s at the wall for 20, 23, and 45 nm particles.

turns of the vane with and without considering the tangential velocity of 2 m/s near the wall, respectively. The area to the right hand side of curve is the particle collection region in which a 23 nm particle will be collected in the cyclone when it starts from this region at the entry of the vane section. Otherwise, it won't be collected. Considering the tangential velocity of 2 m/s on the wall results in a 7% increase in the collection area, and a similar increase in the collection efficiency. The collection areas of three different particle sizes, 20, 23, and 45 nm, are shown in Figure 8b. It shows larger particles have larger collection area than small particles and hence the collection efficiency is also larger.

The experimental particle collection efficiencies and numerical results based on the plug or parab-

oloid flow assumptions are shown in Figure 9. For the operating conditions at P_{in} = 4.3, 5.4, and 6.8 Torr (Figure 9a), and P_{in} = 6 and 7 Torr (Figure 9b). Figure 9a shows that both numerical results agree with the experimental collection efficiencies very well, especially near the cutoff aerodynamic diameter. The error in the numerical prediction of the cutoff aerodynamic diameter by the paraboloid flow assumption is 5.9, 0.7, and 1.5%, for P_{in} = 4.3, 5.4, and 6.8 Torr, respectively. In comparison, the error by the plug flow assumption is larger, which is 16.6, 12.8, and 7.7%,

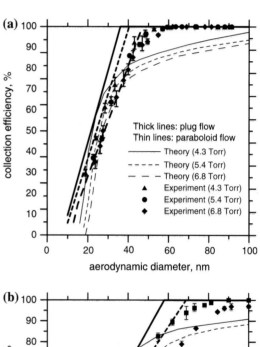

Figure 9. Comparison of numerical collection efficiencis and experimental data. (a) P_{in} = 4.3, 5.4, and 6.8 Torr. (b) P_{in} = 6 and 7 Torr.

for P_{in} = 4.3, 5.4, and 6.8 Torr, respectively. The plug flow assumption over-estimates while the paraboloid flow assumption under-estimates the collection efficiencies when the particle diameter is greater than the cutoff aerodynamic diameter. The maximum difference for the paraboloid flow assumption is around 12%, which is slightly better than the plug flow assumption of 15%.

In Figure 9b, it is seen larger differences occur between the experimental data and theories than in Figure 9a. The error of the cutoff aerodynamic diameter by the paraboloid flow assumptions is 13.9 and 10.1% for P_{in} = 6 and 7 Torr, respectively. In comparison, the error of the plug flow assumption is 25.8 and 11.8 % for P_{in} = 6 and 7 Torr, respectively. The paraboloid assumption predicts the cutoff diameter better than the plug flow assumption.

The differences between the numerical results and the experimental data are mainly due to the over-simplified assumption of either plug flow or paraboloid flow. In fact, the flow is developing near the entry of the vane section. Errors may also be caused by the assumption of a constant velocity of 2 m/s near the wall. In addition, calculating the total migration distance Δr from summing the migration distances of ten segments may also induce errors.

Although the present method is good in predicting the cutoff diameter and collection efficiency, it is not possible to obtain an analytical equation to calculate the cutoff diameter. The following section describes a modified empirical method from Tsai et al. (2004), which can be used to predict the cutoff aerodynamic based on the inlet pressure, flow rate, and cyclone dimension.

Semi-empirical equation for cutoff aerodynamic diameter

In this study, the theoretical derivation of the particle collection efficiency is similar to our previous study (Tsai et al., 2004) but the pressure drop through the vane section is assumed to be linear. The pressure at n^{th} turn of the vane can be defined based on P_{in} (pressure at the cyclone inlet) and P_{out} (pressure at the vane outlet) as

$$P_n = P_{in} - n \times \frac{P_{in} - P_{out}}{n_T}, \tag{13}$$

where the n_T is the total number of turns of the vane and is fixed to 2 based on the present

numerical results. The volumetric flow rate and particle relaxation time will change in the vane section with respect to pressure. The plug flow assumption was used to facilitate the derivation of the theoretical collection efficiency, which is known to under-predict the cutoff diameter. By comparing the experimental cutoff diameters with the theoretical values, we can obtain an empirical factor K to adjust the theoretical values.

Integrating the particle radial velocity and the residence time in the two complete turns, the particle total migration distance (Δr) in the vane section is obtained. The particle collection efficiency is then calculated as in Eq. (9). By setting $\eta = 0.5$ in Eq. (9), the theoretical cutoff aerodynamic diameter, $D_{pa50,theo}$, can be derived as

$$D_{pa50,theo} = \frac{0.11\mu(B-w)\left(r_{max}^2 - r_{min}^2\right)(r_{max} - r_{min})}{\rho_{p0}\lambda_0 r_{min} P_{760}^2}$$
$$\times A, \tag{14}$$

where $A = P_{in} \times P_{out}/Q_0$, which is the important operating parameter of the cyclone as described earlier.

The collection efficiencies of liquid OA particles at 5 different operation conditions are shown in Figure 5 as described in the previous section. The comparison of experimental cutoff diameters at different inlet pressures with the numerical values and the theoretical values by Eq. (14) is shown in Table 1. As expected, the theoretical values are smaller than the experimental results. Therefore, an empirical factor K, which is defined as the ratio of the experimental cutoff size to the theoretical value, is suggested to adjust the theoretical cutoff size. In Table 1, K is shown to be relatively constant with the average and standard deviation of 1.4 and 0.126, respectively. That is, the semi-empirical cutoff aerodynamic diameter can be rewritten as:

$$D_{pa50} = 1.4 D_{pa50,theo}$$
$$= \frac{0.154\mu(B-w)\left(r_{max}^2 - r_{min}^2\right)(r_{max} - r_{min})}{\rho_{p0}\lambda_0 r_{min} P_{760}^2} \times A. \tag{15}$$

The above semi-empirical equation is easy to use and is able to predict the cutoff aerodynamic diameter accurately within 9% of error.

If the collection efficiencies are plotted against $\sqrt{St/St_{50}}$, all the experimental data of five

82

different conditions for OA particles are collapsed into a single curve as shown in Figure 10. In the figure, St is defined as

$$St = \frac{\tau \overline{V}_t}{r_{max} - r_{min}} \qquad (16)$$

and St_{50} corresponds to St at 50% collection efficiency. Referring to Tsai et al. (2004), the slip correction factor used to calculate the particle relaxation time (see Eq. (4)) is given as

$$C(D_{pa}) = \frac{1.695 P_{760}\lambda_0}{\sqrt{P_{in}P_{out}}D_{pa}/2}. \qquad (17)$$

$\sqrt{P_{in}P_{out}}$ is used in this study as the average pressure on which many other parameters depend. Combing Eqs. (4), (11), (12) and (17), St in Eq. (16) is rearranged as

$$St = \frac{0.377 r_{min}\lambda_0\rho_{p0}P_{760}^2}{\mu(r_{max}-r_{min})(r_{max}^2-r_{min}^2)(B-w)A} \times D_{pa}. \qquad (18)$$

It is observed in Figure 10 that all experimental data of particle collection efficiency almost fall on a unique curve, which can be fitted by the following Boltzmann function as

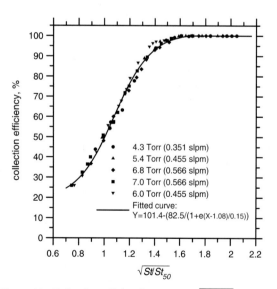

Figure 10. Collection efficiencies *versus* $\sqrt{St/St_{50}}$ for OA particles at different inlet pressures and flow rates.

$$Y = 101.4 - \frac{82.5}{1 + e^{(X-1.08)/0.15}}, X < 1.7, \qquad (19)$$

where Y and X are collection efficiency ($\eta(\%)$) and $\sqrt{St/St_{50}}$, respectively.

Replacing D_{pa} in Eq. (18) by the semi-empirical cutoff aerodynamic diameter, D_{pa50} of Eq. (15), the semi-empirical $\sqrt{St_{50}^*}$ can be calculated as

$$\sqrt{St_{50}^*} = \sqrt{\frac{0.377 r_{min}\lambda_0\rho_{p0}P_{760}^2}{\mu(r_{max}-r_{min})(r_{max}^2-r_{min}^2)(B-w)A} \times D_{pa50}}. \qquad (20)$$

Substituting Eq. (15) into Eq. (20), $\sqrt{St_{50}^*}$ is found to be a fixed value of 0.241. The experimental collection efficiency *versus* $\sqrt{St}$ (calculated from Eq. (18)) for all five test conditions was plotted (not shown here), and the experimental $\sqrt{St_{50}}$ was calculated to be 0.256, 0.240, 0.222, 0.226 and 0.240 for P_{in} = 4.3, 5.4, 6.8, 7.0 and 6.0 Torr, respectively. The difference between the experimental $\sqrt{St_{50}}$ and the semi-empirical $\sqrt{St_{50}^*}$, or 0.241, is −6.2, 0.4, 7.9, 1.5 and 0.4 % for P_{in} = 4.3, 5.4, 6.8, 7.0 and 6.0 Torr, respectively. That is, for the present axial flow cyclone for removing nanoparticles, the design value for the square root of the cutoff Stokes number is

$$\sqrt{St_{50}^*} = 0.241. \qquad (21)$$

This value is much smaller than the $\sqrt{St_{50}}$ of the circular jet and rectangular jet impactor, which is 0.49 and 0.77, respectively (Hinds, 1999).

Conclusions

In this study, the axial flow cyclone of Tsai et al. (2004) was further tested for the collection efficiencies of solid NaCl and liquid OA nanoparticles in the diameter from 12 to 100 nm at low-pressure conditions (4.3, 5.4, 6.0, 6.8, and 7.0 Torrs). The objective was to find whether solid particle bounce would influence the collection efficiencies. The axial flow cyclone was shown to be able to remove NaCl and OA nanoparticles below 50 nm efficiently. The smallest cutoff aerodynamic diameters of OA and NaCl particles were found to be 21.7 (4.3 Torr, 0.351 lpm) and 21.2 (5.4 Torr, 0.454 lpm), respectively. Differences in the collection efficiencies of liquid and solid particles were within 10% indicating that the effect of solid particle bounce on the collection efficiency was not very obvious. The test for the solid particle loading in the cyclone further indicated a soiled cyclone wall did not change the collection efficiency too much (< 10%).

In order to predict the collection efficiency and cutoff diameter accurately, a 3-D numerical simulation was conducted to obtain the pressure distribution and gas velocity fields in the cyclone. Results showed that the tangential flow developed quickly into paraboloid flow profile near the end of the first turn of the vane section. Total migration distance was calculated based on the local tangential flow velocity and pressure to obtain the numerical collection efficiency.

The comparison of the experimental particle collection efficiencies and cutoff diameters with the numerical simulations showed that the Paraboloid tangential flow assumption yielded better results than the plug flow assumption, with the maximum error of 15% for the collection efficiencies and 13.9% for the cutoff diameters, respectively.

Based on the simulated pressure and tangential flow fields, a modified theoretical method from Tsai et al. (2004) was proposed. The semi-empirical equation was then developed to predict the cutoff aerodynamic diameter and the cutoff Stokes number accurately within 9% and 8% of error, respectively. Based on the semi-empirical cutoff aerodynamic diameter, the design value of the square root of the cutoff Stokes number, $\sqrt{St_{50}^*}$, was calculated and found to be a constant value of 0.241.

Acknowledgement

Authors would like to thank for the financial support of this project by Taiwan National Science Council (NSC 94-2211-E-009-001).

References

Biswas P. & R.J. Flagan, 1988. Particle trap impactor. Aerosol Sci. 19, 113–121.

Blachman M.W. & M. Lippmann, 1974. Performance characteristics of the multicyclone aerosol sampler. Am. Ind. Hyg. Assoc. J. 35, 311–326.

Boysan F., B.C.R. Ewan, J. Swithenbank & W.H. Ayers, 1983. Experimental and theoretical studies of cyclone aeparator aerodynamics. IChemeE. Symp. Series 69, 305–320.

Harwood R. & M. Slack, 2002. CFD analysis of a cyclone. QNET-CFD Network Newsletter. 1, 25–27.

Hinds W.C., 1999 Aerosol Technology. New York: Wiley 126.

Hoekstra A.J., J.J. Derksen & H.E.A. Van Den Akker, 1999. An experimental and numerical study of turbulent swirling flow in gas cyclones. Chem. Eng. Sci. 54, 2055–2065.

Hsu Y.D., H.M. Chein, T.M. Chen & C.J. Tsai, 2005. Axial flow cyclone for segregation and collection of ultrafine particle: Theoretical and experimental study. Environ. Sci. Technol. 39, 1299–1308.

Liu B.Y.H. & K.L. Rubow, 1984. A new axial flow cascade cyclone for size characterization of airborne particulate matter. In: Liu B.Y.H., Pui D.Y., & Fissan H.J. ed. Aerosols. Elsevier, Amsterdam, pp. 115–118.

Maynard A.D., 2000. A simple model of axial flow cyclone performance under laminar flow conditions. J. Aerosol Sci. 31, 151–167.

Patankar S.V., 1980 Numerical Heat Transfer and Fluid Flow. Washington: Hemisphere Publishing Co.

Schmidt S. & F. Thiele, 2002. Comparison of numerical methods applied to the flow over wall-mounted cubes. Inter. J. Heat & Fluid Flow. 23, 330–339.

Schmidt S., H.M. Blackburn, M. Rudman & I. Sutalo, 2003. Simulation of turbulent flow in a cyclonic separator. 3rd International conference on CFD in the Minerals and Process Industries CSIRO, Melbourne, Australia, 10–12 December 2003, p. 365–369.

Tsai C.J., H.G. Shiau, K.C. Lin & T.S. Shih, 1999. Effect of deposited particles and particle charge on the penetration of small sampling cyclones. J. Aerosol Sci. 30, 313–323.

Tsai C.J. & T.I. Lin, 2000. Particle collection efficiency of different impactor designs. Sep. Sci. Technol. 35, 2639–2650.

Tsai C.J. & Y.H. Cheng, 1995. Solid particle collection characteristics on impaction surfaces of different designs. Aerosol Sci. Technol. 23, 96–106.

Tsai C.J., D.R. Chen, H.M. Chein, S.C. Chen, J.L. Roth, Y.D. Hsu, W. Li & P. Biswas, 2004. Theoretical and experimental study of an axial flow cyclone for fine particle removal in vacuum conditions. J. Aerosol Sci. 35, 1105–1118.

Vaughan N.P., 1988. Construction and testing of an axial flow cyclone pre-separator. J. Aerosol Sci. 19, 295–305.

Weiss Z., P. Martinec & J. Vitek, 1987. Vlastnosti Dulnibo Prachu A Zaklady Protiprasne Techniky. Prague: SNTL.

Xiang R.B. & K.W. Lee, 2004. The flow pattern in cyclones with different cone dimensions and its effect on separation efficiency. Abstract, EAC 2004, Budapest, Hungary, Sep. 6–10, p. 289–290.

Zhu Y. & K.W. Lee, 1999. Experimental study on small cyclones operating at high flowrates. J. Aerosol Sci. 30, 1303–1315.

Journal of Nanoparticle Research (2007) 9:85–92
DOI 10.1007/s11051-006-9178-2

Special issue: Nanoparticles and Occupational Health

Measuring particle size-dependent physicochemical structure in airborne single walled carbon nanotube agglomerates

Andrew D. Maynard[1,*], Bon Ki Ku[2], Mark Emery[3], Mark Stolzenburg[3] and Peter H. McMurry[3]
[1]*Woodrow Wilson International Center for Scholars, Project on Emerging Nanotechnologies, One Woodrow Wilson Plaza, 1300 Pennsylvania Avenue NW, Washington, DC, 20004, USA;* [2]*Centers for Disease Control and Prevention, National Institute for Occupational Safety and Health, 4676 Columbia Parkway, Cincinnati, OH, 45226, USA;* [3]*Mechanical Engineering Department, University of Minnesota, 111 Church Street, Minneapolis, MN, 55455, USA;* *Author for correspondence (Tel.: +1-202-691-4311; Fax: +1-202-691-4001; E-mail: andrew.maynard@wilsoncenter.org)*

Received 14 August 2006; accepted in revised form 5 September 2006

Key words: aerosol, carbon nanotubes, differential mobility analysis, aerosol particle mass monitor, composite nanoparticles, occupational health

Abstract

As-produced single-walled carbon nanotube (SWCNT) material is a complex matrix of carbon nanotubes, bundles of nanotubes (nanoropes), non-tubular carbon and metal catalyst nanoparticles. The pulmonary toxicity of material released during manufacture and handling will depend on the partitioning and arrangement of these components within airborne particles. To probe the physicochemical structure of airborne SWCNT aggregates, a new technique was developed and applied to aerosolized as-produced material. Differential Mobility Analysis-classified aggregates were analyzed using an Aerosol Particle Mass Monitor, and a structural parameter Γ (proportional to the square of particle mobility diameter, divided by APM voltage) derived. Using information on the constituent components of the SWCNT, modal values of Γ were estimated for specific particle compositions and structures, and compared against measured values. Measured modal values of Γ for 150 nm mobility diameter aggregates suggested they were primarily composed of non-tubular carbon from one batch of material, and thin nanoropes from a second batch of material – these findings were confirmed using Transmission Electron Microscopy. Measured modal values of Γ for 31 nm mobility diameter aggregates indicated that they were comprised predominantly of thin carbon nanoropes with associated nanometer-diameter metal catalyst particles; there was no indication that either catalyst particles or non-tubular carbon particles were being preferentially released into the air. These results indicate that the physicochemistry of aerosol particles released while handling as-produced SWCNT may vary significantly by particle size and production batch, and that evaluations of potential health hazards need to account for this.

Introduction

Single walled carbon nanotubes (SWCNT) represent a carbon-based material that shows unique physical and chemical properties. As applications of the material continue to be explored, possible implications to human health following exposure have been raised (Shvedova et al., 2003, 2005; Lam et al., 2004; Maynard et al., 2004; Warheit et al., 2004). The potential health hazards of single walled carbon nanotubes (SWCNT) remain poorly understood. However, it is probable that the unusual physicochemical properties underpinning their value as a new engineered nanomaterial – including high tensile strength and structure-dependent electrical conductivity – will influence their toxicity, as well as determine deposition and transportation within the body. Early studies have shown SWCNT are cytotoxic and lead to oxidative stress *in vitro* (Shvedova et al., 2003), as well as indicating pro-inflammatory and fibrogenic responses in rodent lungs (Lam et al., 2004; Warheit et al., 2004). In particular, biological response has been associated with the nanoscale structure and composition of SWCNT particles (Shvedova et al., 2003, 2005; Warheit et al., 2004). Although aerosol generation rates have been shown to be low during handling (Maynard et al., 2004), published data to date indicate that inhaled airborne material may present a pulmonary hazard.

As-produced SWCNT material is a complex matrix of carbon nanotubes, bundles of nanotubes (nanoropes), non-tubular carbon and metal catalyst nanoparticles (Bronikowski et al., 2001). The pulmonary toxicity of as-produced SWCNT released during manufacture and handling will depend on the partitioning and arrangement of these components within airborne particles. However, physicochemical characterization of airborne particles at the nanoscale presents unique challenges. Here we present a new technique for probing the nanostructure of airborne particles, and use it to show significant differences in nanostructure between 150 nm mobility diameter and 31 nm mobility diameter aerosol particles released from as-produced SWCNT.

Background

It has previously been shown that as-produced SWCNT material generated using the HiPCO™ process can release particles into the air when agitated (Maynard et al., 2004). Controlled agitation of material from a SWCNT manufacturer using the HiPCO™ process demonstrated a bimodal aerosol, with modal count median mobility diameters of approximately 400 nm, and

below 10 nm (Maynard et al., 2004). Although the generation rates observed during laboratory and small-scale production conditions were very low, potential health risk following inhalation will depend on the toxicity of the airborne particles in combination with exposure.

Transmission Electron Microscopy (TEM) of the same as-produced SWCNT material confirmed that it comprised of four distinct components: carbon-coated iron-rich catalyst particles approximately 5 nm in diameter, discrete single walled carbon nanotubes, ordered bundles of single walled carbon nanotubes typically between 5 and 50 nm in diameter and non-tubular carbon material (Figure 1). The mass-fraction of iron was previously measured at 30% using inductively coupled plasma – mass spectrometry (Maynard et al., 2004). The fraction of non-tubular carbon was not quantified, although TEM analysis indicated relatively little to be present in the bulk material.

Each of these components is likely to represent a different toxicity within the lungs. The US National Institute for Occupational Safety and Health (NIOSH) Recommended Exposure Limits (REL) for carbon black (assumed to be analogous to the non-tubular carbon) and iron oxide fume

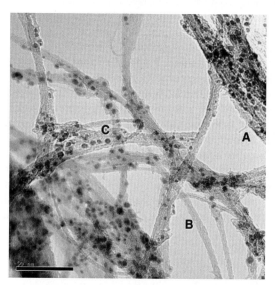

Figure 1. TEM image of as-produced SWCNT (second batch of material). Three of the four principle components are identified: A: discrete single walled carbon nanotubes; B: bundles of carbon nanotubes (nanoropes); C: iron catalyst particles. The non-tubular carbon component is not visible in this micrograph.

are 3.5 mg/m^3 and 5 mg/m^3 respectively. There have been indications that SWCNT may be as toxic as crystalline quartz in the lungs (Lam et al., 2004) – which has an REL of 0.05 mg/m^3 – and Shvedova et al. have indicated that the presence of nanometer-diameter iron particles within as-produced SWCNT lead to Fenton-type reactions and the release of reactive oxygen species (Shvedova et al., 2003). An unusual inflammatory and fibrogenic response to SWCNT aspirated into the lungs of mice has also been observed (Shvedova et al., 2005).

In addition to an inhalation hazard determinant on the composition of SWCNT-derived airborne particles, the physical structure of the generated particles will possibly contribute to their pulmonary toxicity. Shvedova et al. (2005) identified distinct responses in the proximal and distal regions of mice lungs aspirated with SWCNT, and speculated that the regional responses were associated with different agglomerate structures.

A recent review of engineered nanomaterial toxicity screening tests emphasized the importance of structure as well as chemistry in determining potential hazard, and stressed the need to fully characterize the physicochemical nature of materials under study (Oberdörster et al., 2005). In the case of particles released from as-produced SWCNT, possible variations in particle structure and composition place stringent requirements on characterization, which push the bounds of current analytical techniques. When agitated under controlled conditions, 150 nm mobility diameter particles were released from as-produced SWCNT at a sufficient rate to allow physicochemical characterization using TEM. However, given low aerosol generation rates and material availability, it is estimated that TEM analysis of particles in the sub−35 nm diameter mode would require unfeasibly long sample collection times.

Method

To explore the nature of smaller aerosol particles released from the bulk material, we have developed a new approach to probing the physical structure of airborne nanoparticles by combining differential mobility analysis (DMA) with Aerosol Particle Mass analysis (APM). These two techniques have previously been used together to explore the structural properties of aerosols (McMurry et al., 2002; Park et al., 2004b). Here, we define the structural parameter Γ, and use it to estimate the composition and structure of sampled airborne particles.

Differential mobility analysis enables airborne particles with a specific electrical mobility diameter to be differentially selected from an aerosol, and is capable of producing monodisperse aerosols with modal diameters from a few nanometers upwards (Knutson & Whitby, 1975). For open-structured and compact particles with Knudsen numbers greater than 1 (Kn, defined as the mean free path of the suspension gas divided by particle radius), electrical mobility diameter is comparable to the diameter of a sphere with the same projected area (Rogak et al., 1993; Park et al., 2004a; Ku & Maynard, 2005). Mobility diameter can therefore be used to derive particle surface area for a range of particle morphologies. In general, particles differentially classified using the DMA will have the same 'active' surface area – defined as the surface area associated with particle-molecule interactions that determine drag (Keller et al., 2001).

APM analysis is a recently developed method for measuring the mass of discrete singly-charged airborne particles (Ehara et al., 1996; McMurry et al., 2002). Particles of a known mobility diameter and charge are transported axially through a narrow annular gap separating two rotating concentric cylindrical electrodes. The cylinders and particles rotate about the axis at the same angular speed. High mass particles deposit on the outer electrode, while low-mass particles deposit on the inner electrode due to electrophoresis: Only those particles where centrifugal force is precisely balanced by the inward electrostatic force traverse the device, where they are counted using a condensation particle counter. By varying the voltage between the electrodes, particles with a specific charge to mass ratio are differentially sampled. For instance, at 1000 rpm and 500 volts, particles with a modal mass of 3.95×10^{-17} kg are differentially selected, corresponding to 423 nm diameter spherical particles with a density of 1000 kg/m^3. Instrument resolution is influenced by rotational velocity, particle size and sampling flow rate (Ehara et al., 1996). McMurry et al. (2002) have shown the APM to have a relatively broad transfer function using polystyrene latex spheres, although modal values of particle number versus APM voltage enabled the

Table 1. Estimated values of Γ (Γ$_{est}$) as a function of particle composition and structure, for a variety of possible particle morphologies and compositions

ID[a]	Mobility Diameter (nm)	Description[b]	Γ$_{est}$ (m^2/g)[c]
A	150	Compact non-tubular carbon particles	20
B	150	Compact non-tubular carbon particles with 30% Fe by mass	11
C	150	Compact single walled carbon nanotubes with 30% Fe by mass	12
D	150	Open agglomerate of single walled carbon nanotube with 30% Fe by mass	860
E	150	Open agglomerate of 15 nm diameter nanoropes with 30% Fe by mass	81
F	150	Open agglomerate of 5 nm diameter Fe particles	150
G	31	Compact non-tubular carbon particles	97
H	31	Compact non-tubular carbon particles with 30% Fe by mass	51
I	31	Compact single walled carbon nanotubes with 30% Fe by mass	58
J	31	Open agglomerate of single walled carbon nanotube with 30% Fe by mass	860
K	31	Open agglomerate of 5 nm diameter nanoropes with 30% Fe by mass	240
L	31	Open agglomerate of 5 nm diameter Fe particles	150

Values of Γ$_{est}$ are based on estimates of particle mass and surface area for a given composition and structure
[a] Refer to Figures 2 and 3.
[b] Compact particles are assumed to have a minimum surface area. Open agglomerates are assumed to have a surface area determined by the dimensions of the component structures.
[c] Estimated assuming a density for carbon nanotubes and nanoropes of 1.4 g/cm^3, a density of 7.9 g/cm^3 for iron nanoparticles and a density of 2 g/cm^3 for non-tubular carbon.

mass of spherical monodisperse aerosol particles to be measured to within 5%.

In order to extend the operational range to sub-100 nm diameter particles with low densities, the APM was operated at up to 4100 rpm and 0.5 l/min sampling rate, with differential voltages between the electrodes of less than 2 volts. This revised configuration enabled differential separation of

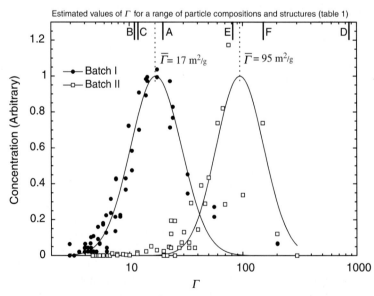

Figure 2. Measured values of Γ for 150 nm mobility diameter as-produced SWCNT-derived particles. Estimated values of Γ (Γ$_{est}$) for particles with specific physicochemical characteristics (Table 1) are shown for comparison along the upper x-axis. Particles from batch I have a (lognormal fit) value of Γ̄ of 17 m^2/g, while the equivalent parameter for particles from batch II is 95 m^2/g. Comparing values of Γ̄ with Table 1, particles from batch I are indicated as being predominantly compact particles, possibly with a significant non-tubular carbon component. Particles from batch II have a distinctly higher value of Γ̄, indicative of open agglomerates of small-diameter nanoropes (Table 1).

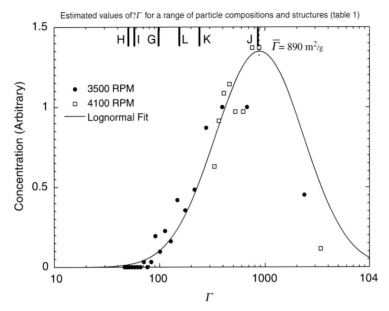

Figure 3. Measured values of Γ for 31 nm mobility diameter as-produced SWCNT-derived particles (batch I material). Estimated values of Γ (Γ_{est}) for particles with specific physicochemical characteristics (Table 1) are shown for comparison along the upper x-axis. Based on a lognormal fit, the distribution has a value of $\bar{\Gamma}$ of 890 m²/g, although the quality of the lognormal fit at higher values is not good. The relatively high value of $\bar{\Gamma}$ suggests that the particles had an open structure dominated by discrete carbon nanotubes and nanometer-diameter nanoropes (Table 1).

particles with a mass below 2×10^{-21} kg (corresponding to a 31 nm diameter spherical particle with a density of 600 kg/m³).

Particles with a selected mobility diameter (and therefore active surface area) were classified by mass in the APM. APM spectra for a given particle mobility diameter were used to calculate Γ, defined as

$$\Gamma = \pi \bar{d}_m^2 \frac{\omega^2 \bar{r}}{qE} \qquad (1)$$

where $\bar{d}_m$ is the modal particle mobility diameter selected by the DMA, q is the particle charge, $\bar{r}$ is the gap separating the concentric cylinders within the APM, ω is the rotational velocity of the APM electrodes and E is the electric field between the two electrodes.

Γ represents modal particle surface area (defined by the DMA), divided by modal particle mass (defined by the APM), and has units of surface area per unit mass. Plots of particle number concentration versus Γ represent a parameter closely related to particle specific surface area, convoluted by the DMA and APM transfer functions. Without de-convolution (particularly for the APM transfer function) Γ can not provide detailed

structural information on aerosol particles. However, by varying E and/or ω for a given $\bar{d}_m$ and measuring particle number concentration, the modal value of Γ ($\bar{\Gamma}$) for a given mobility diameter may be estimated. Comparing $\bar{\Gamma}$ with calculated values of specific surface area for possible particle structures and compositions enables some insight to be gained into the physicochemical nature of the aerosol under study.

Measurement of Γ for SWCNT-derived particles with mobility diameters of 31 nm and 150 nm were made, representing particles within the two aerosol size distribution modes observed by Maynard et al. In all cases, SWCNT material was aerosolized using the method previously described (Maynard et al., 2004). Characterization of the larger diameter particles was validated using TEM. 150 nm mobility diameter particles from two batches of SWCNT produced using the same process were analyzed: A similar comparison was not possible for 31 nm diameter particles, due to excessively low generation rates from the second batch of material. Replicate measurements were made over a range of APM rotational velocities. As the aerosol generation rate varied with time, measurements made under the same conditions were combined,

90

and then normalized for concentration within overlapping ranges of the parameter Γ. Modal values of Γ ($\bar{\Gamma}$) were estimated by fitting data using a lognormal function. Measured values of Γ were compared with estimates (Γ_{est}) associated with various particle compositions and morphologies (Table 1, Figures 2, 3).

Results and discussion

Modal values of Γ for 150 nm diameter particles differed between the two batches of material tested (Figure 2). Comparison between $\bar{\Gamma}$ for batch I and Γ_{est} (Table 1) indicated that compact particles of non-tubular carbon material were being preferentially released into the air. TEM analysis confirmed this, although sampled 100 nm mobility diameter particles had a projected area-equivalent diameter closer to 400 nm (Figure 4). Measuring projected

area as a function of sample holder tilt within the TEM confirmed that these particles were planar, suggesting that particles within the DMA either aligned with the electrostatic field or tumbled, thus reducing their apparent mobility diameter.

DMA/APM analysis on batch II of SWCNT material indicated that the particles being released were dominated by a nanostructure which was relatively open (high values of Γ). Comparison of $\bar{\Gamma}$ with estimated values (Table 1) for a range of physicochemical structures, suggested that the particles being released were associated with small-diameter nanoropes or combinations of nanoropes and other nanostructured components. This was confirmed in TEM micrographs of sampled 150 nm mobility diameter particles (Figure 5). Size distribution measurements made on the batch I material confirmed previously published size distribution (Maynard et al., 2004). However, similar measurements on the batch II material

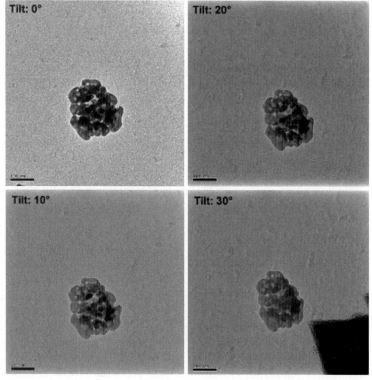

Figure 4. Micrographs of a representative 100 nm mobility diameter particle from batch I material, imaged over a number of specimen stage tilts. The particle was collected onto a lacey carbon support film using a thermophoretic sampler (Maynard, 1995). Although the projected area is indicative of particles with a mobility diameter larger than 100 nm, decreasing projected area with increasing tilt angle shows the particles to be planar. The particle is compact, and appears to be predominantly non-tubular carbon, in agreement with the physicochemical structure inferred from the modal value of Γ (Figure 2).

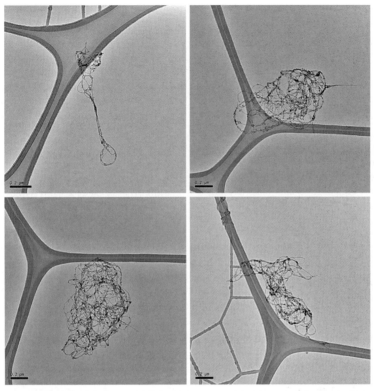

Figure 5. Micrographs of representative 150 nm mobility diameter particles from batch II material. Particles were collected onto a carbon support grid using a thermpohoretic aerosol sampler (Maynard 1995). The particles have an open structure comprising predominantly of nanometer-diameter nanoropes and Fe catalyst particles, in agreement with the physicochemical structure inferred from the data in Figure 2.

were very different: The lower-diameter mode was not well developed, and the upper-diameter mode occurred well above the upper size limit of the DMA (approximately 400 nm in this case). Although the SWCNT material for both batches was produced by the same manufacturer under similar conditions, our results indicate that subtle changes in physicochemical properties may have profound effects on hazard potential.

The estimated value of $\bar{\Gamma}$ for 31 nm mobility diameter particles from batch I material indicated a strong dependence on nanostructure (Figure 3). The analysis method was operating at its limits above approximately 800 m^2/g, and data within this region need to be treated with care. The lognormal fit used to estimate $\bar{\Gamma}$ clearly deviates systematically from the data above $\Gamma = 500$ m^2/g, although the quality of the data are not sufficient to indicate whether the deviation is significant or not. However, comparison with estimated value of $\bar{\Gamma}$

clearly indicates particles with an open structure dominated by nanometer-diameter nanotubes. There is little evidence that compact carbonaceous particles were released.

Summary

Associations between structure and the properties of engineered nanomaterials challenge current approaches to evaluating hazard potential, and in many cases indicate new methods and approaches will be needed for characterizing materials. Using a new aerosol analysis technique, the size and source material-dependent physical and chemical complexity of single walled carbon nanotube aerosols has been demonstrated. Our results suggest that the physicochemistry of aerosol particles released while handling as-produced SWCNT may vary significantly by particle size and production batch, and that evaluation of potential health hazard needs to account for this.

References

Bronikowski M.J., P.A. Willis, D.T. Colbert, K.A. Smith & R.E. Smalley, 2001. Gas-phase production of carbon single-walled nanotubes from carbon monoxide via the HiPCO® process: A parametric study. J. Vac. Sci. Technol. A.-Vac. Surf. Films 19(4), 1800–1805.

Ehara K., C. Hagwood & K.J. Coakley, 1996. Novel method to classify aerosol particles according to their mass-to-charge ratio – aerosol particle mass analyzer. J. Aerosol Sci. 27(2), 217–234.

Keller A., M. Fierz, K. Siegmann, H.C. Siegmann & A Filippov, 2001. Surface science with nanosized particles in a carrier gas. J. Vacuum Sci. Technol. Vacuum Surf. Films 19(1), 1–8.

Knutson E.O. & K.T. Whitby, 1975. Aerosol classification by electrical mobility: Apparatus, theory, and applications. J. Aerosol Sci. 6, 443–451.

Ku B.K. & A.D Maynard, 2005. Comparing aerosol surface-area measurement of monodisperse ultrafine silver agglomerates using mobility analysis, transmission electron microscopy and diffusion charging. J. Aerosol Sci. 36(9), 1108–1124.

Lam C.-W., J.T. James, R. McCluskey & R.L. Hunter, 2004. Pulmonary toxicity of single-wall carbon nanotubes in mice 7 and 90 days after intratracheal instillation. Toxicol. Sci. 77, 126–134.

Maynard A.D., 1995. The development of a new thermophoretic precipitator for scanning-transmission electron-microscope analysis of ultrafine aerosol-particles. Aerosol Sci. Technol. 23(4), 521–533.

Maynard A.D., P.A. Baron, M. Foley, A.A. Shvedova, E.R. Kisin & V. Castranova, 2004. Exposure to carbon nanotube material: Aerosol release during the handling of unrefined single walled carbon nanotube material. J. Toxicol. Environ. Health 67(1), 87–107.

McMurry P.H., X. Wang, K. Park & K. Ehara, 2002. The relationship between mass and mobility for atmospheric particles: A new technique for measuring particle density. Aerosol Sci. Technol. 36(2), 227–238.

Oberdörster G., A. Maynard, K. Donaldson, V. Castranova, J. Fitzpatrick, K. Ausman, J. Carter, B. Karn, W. Kreyling, D. Lai, S. Olin, N. Monteiro-Riviere, D. Warheit & H. Yang, 2005. Principles for characterizing the potential human health effects from exposure to nanomaterials: elements of a screening strategy. Part. Fiber Toxicol. 2(8): doi:10.1186/1743-8977-2-8.

Park K., D.B. Kittelson & P.H. McMurry, 2004a. Structural properties of diesel exhaust particles measured by transmission electron microscopy (TEM): Relationships to particle mass and mobility. Aerosol Sci. Tech. 38(9), 881–889.

Park K., D.B. Kittelson, M.R. Zachariah & P.H. McMurry, 2004b. Measurement of inherent material density of nanoparticle agglomerates. J. Nanopart. Res. 62(2), 267–272.

Rogak S.N., R.C. Flagan & H.V. Nguyen, 1993. The mobility and structure of aerosol agglomerates. Aerosol Sci. Technol. 18(1), 25–47.

Shvedova A.A., E.R. Kisin, R. Mercer, A.R. Murray, V.J. Johnson, A.I. Potapovich, Y.Y. Tyurina, O. Gorelik, S. Arepalli, D. Schwegler-Berry, A.F. Hubbs, J. Antonini, D.E. Evans, B.K. Ku, D. Ramsey, A. Maynard, V.E. Kagan, V. Castranova & P Baron, 2005. Unusual inflammatory and fibrogenic pulmonary responses to single-walled carbon nanotubes in mice. Am. J. Physiol.-Lung Cell. Mol. Physiol. 289, 698–708.

Shvedova A.A., E.R. Kisin, A.R. Murray, V.Z. Gandelsman, A.D. Maynard, P.A. Baron & V. Castranova, 2003. Exposure to carbon nanotube material: Assessment of the biological effects of nanotube materials using human keratinocyte cells. J. Toxicol. Environ. Health 66(20), 1909–1926.

Warheit D.B., B.R. Laurence, K.L. Reed, D.H. Roach, G.A.M. Reynolds & T.R. Webb, 2004. Comparative pulmonary toxicity assessment of single-wall carbon nanotubes in rats. Toxicol. Sci. 77, 117–125.

Journal of Nanoparticle Research (2007) 9:93–107
DOI 10.1007/s11051-006-9179-1

Special issue: Nanoparticles and Occupational Health

A comparison of two nano-sized particle air filtration tests in the diameter range of 10 to 400 nanometers

Daniel A. Japuntich[1,*], Luke M. Franklin[2], David Y. Pui[2], Thomas H. Kuehn[2], Seong Chan Kim[2] and Andrew S. Viner[1]
[1]3M Company, Building 201-2E–04, Saint Paul, MN, 55144, USA; [2]Institute of Technology, University of Minnesota, 1100 Mechanical Engineering, 111 Church Street SE, 3101, Minneapolis, MN, 55455-0111, USA; *Author for correspondence (Tel.: + 1-651-736-9041; Fax: + 1-651-737-2590; E-mail: dajapuntich5@ mmm.com)

Received 1 September 2006; accepted in revised form 5 September 2006

Key words: filter test, nanoparticles, condensation particle counters, differential mobility analyzer (DMA), scanning mobility particle sizer (SMPS), occupational health

Abstract

Two different air filter test methodologies are discussed and compared for challenges in the nano-sized particle range of 10–400 nm. Included in the discussion are test procedure development, factors affecting variability and comparisons between results from the tests. One test system which gives a discrete penetration for a given particle size is the TSI 8160 Automated Filter tester (updated and commercially available now as the TSI 3160) manufactured by the TSI, Inc., Shoreview, MN. Another filter test system was developed utilizing a Scanning Mobility Particle Sizer (SMPS) to sample the particle size distributions downstream and upstream of an air filter to obtain a continuous percent filter penetration versus particle size curve. Filtration test results are shown for fiberglass filter paper of intermediate filtration efficiency. Test variables affecting the results of the TSI 8160 for NaCl and dioctyl phthalate (DOP) particles are discussed, including condensation particle counter stability and the sizing of the selected particle challenges. Filter testing using a TSI 3936 SMPS sampling upstream and downstream of a filter is also shown with a discussion of test variables and the need for proper SMPS volume purging and filter penetration correction procedure. For both tests, the penetration versus particle size curves for the filter media studied follow the theoretical Brownian capture model of decreasing penetration with decreasing particle diameter down to 10 nm with no deviation. From these findings, the authors can say with reasonable confidence that there is no evidence of particle thermal rebound in the size range.

Introduction

Nano-sized particles are often defined as particles in the size range of less than 100 nanometers (nm). Particles in this range fall into two categories, nanoparticles and ultra fines. Nanoparticles are the end products of applying nano-technology to various materials for industrial applications and products. Ultra fines are created as the result of a synthetic or natural process, and tend to be formed through nucleation, gas to particle reactions, or evaporation. Much progress is being made in industrial and academic research with respect to the manufacture of nanoparticles from a wide range of materials to determine their unique properties and potential for industrial applications. Examples of industrial processes that may produce ultra fines or nano-size particles include flame pyrolysis, thermal spraying and coating, and welding. Particles in the nanometer size range are also produced in large quantities from diesel engines and from domestic activities such as gas cooking. Nanometer sized particles are also found in the atmosphere where they originate from combustion sources (traffic, forest fires), volcanic activity, and from atmospheric gas to particle conversion processes such as photochemically driven nucleation (Aitken et al., 2004).

Engineered nanoparticles are being used in electronic, magnetic and optoelectronic, biomedical, pharmaceutical, cosmetic, energy and catalytic applications. The ability to control a combination of unique material properties at a specific nano-size range has led to speculation on a number of fronts, from measuring and filtering nano-size particles to potential health effects.

If airborne nano-size particles need to be controlled, one method is filtration, and questions have been raised regarding the capturing mechanisms and possible deviation from the classical single-fiber-efficiency theory. In addition to performance questions, research is underway to understand if there are any unanticipated adverse health effects of materials at the nano-sized range. Current exposure indices are based on the mass concentration of the particle, and research is underway to determine if this is the most appropriate metric to represent the health effects of nanoparticles and ultra fine materials. The nano-size particle debate points out that the concentration might be small in terms of mass, quite large in surface area, and even greater in terms of particle numbers. Occupational hygienists need to collect personal exposure samples that represent a biologically relevant fraction, and currently not enough knowledge exists to determine how these nano-sized particles should be measured and assessed. Although understanding the health effects of nano-sized aerosols and the ability to measure those fractions will likely be a long process, it is important that research begin on understanding the performance of filtration media challenged with nano-sized particles, and to determine if the capture of nano-sized particles deviates from current filtration theory.

As guiding references for this filtration study, testing standards and protocols for filter penetration versus particle size have been put forth by the American Society of Heating Refrigeration and Air-Conditioning Engineers (ASHRAE) and the American Society for Testing and Materials (ASTM). ASHRAE Standard 52.2 (1999) "Method of Testing General Ventilation Air-Cleaning Devices for Removal Efficiency by Particle Size" outlines specific requirements for the evaluation of penetration through filter media. ASTM Standards F778-88 (2001) "Standard Methods for Gas Flow Resistance Testing of Filtration Media" and F2519-05 (2005) "Standard Test Method for Grease Particle Capture Efficiency of Commercial Kitchen Filters and Extractors" also outline similar requirements.

The theory of aerosol filtration through fibrous filter media is well developed, as shown in Kirsch and Stechkina (1978), Brown (1993) and Hinds (1999). In the nano-sized particle range, the dominant mechanism for particle deposition on filter fibers is Brownian diffusion, which for aerosol flow through a filter medium is described by analytical relationships with the dimensionless fiber Peclet number ($Pe = d_f U_o/D$, where d_f is the fiber diameter, U_o is the filter face velocity and D is the particle diffusion coefficient). For uncharged fibrous filters the theory shows that as particle size decreases below 100 nm, filter percent penetration continuously decreases because of increasing diffusional deposition. Kirsch (2003) reviewed the theory of particle filtration in the nano-sized particle ranges and presented modeling in the Brownian convective diffusion regime for any Peclet number, including those less than one for particles as small as 2 nm, and this work could be used to check future test results in this very low range of size.

Lately, research interest has increased in the filtration of particle sizes below 30 nm. Theoretically, for very small particles less than 10 nm (Wang & Kasper, 1991), a percentage of solid particles may develop enough thermal energy from gas diffusional collisions to bounce from fibers or "rebound" just as if they were gas molecules. This has been studied empirically for screen and wire diffusion batteries with thermal rebound occurring at no greater than 3 nm by Skaptsov et al. (1996), Cheng and Yeh (1980), and Wang (1996). Heim et al. (2005) states that for particles as low as 2.5 nm, no measurable deviation occurs from the classical single-fiber-efficiency theory of air filtration. For silver nanoparticles, Kim et al. (2005) have confirmed this claim, showing fibrous filter testing results that substantiate a continuous decrease in filter penetration with decreasing particle diameter down to 3 nm with no evidence of thermal rebound.

One method of obtaining the penetration of particles through an air filter is to measure it at a set particle size using a monodisperse aerosol challenge. This involves producing a stable small particle size distribution, which is then electrically classified into a narrow (monodisperse) particle size distribution with a differential mobility analyzer, electrostatically neutralized and finally used as a filter challenge. The filter upstream and downstream count concentrations may be measured using condensation particle counters, and the filter penetration may be calculated as the ratio of the downstream to the upstream concentrations for each discrete particle size challenge. The challenge aerosol size may be increased or decreased to give a range of discrete filter penetrations. A good example of such a system is shown in Dhaniyala and Lui (1999). The TSI 3160 Automated Filter Tester (an updated version of the TSI 8160) is a commercially available example of this system. Lifshutz and Pierce (1996) described an extensive study of penetration results of flat filter media with respect to the "most penetrating particle size" (MPPS, see Lee and Lui (1980)) using the TSI 8160 Automated Filter Tester, comparing the results to data from the TSI 8160 CertiTest Automated Filter Tester (a replacement for the old ATI Q127 DOP penetrometer high efficiency filter test). Franklin (2005) conducted an in-depth study of variables affecting the performance of a TSI 8160 and an outline of a methodology to reduce them, and this is a reference for a good calibration protocol for the TSI 8160/8130.

Another method of filter penetration testing for nano-sized particles is to produce a stable, small particle aerosol with a wide size distribution, electrostatically neutralize the aerosol and use it as a direct filter challenge. The concentration of various particle sizes upstream and downstream of the filter may be measured with an aerosol spectrometer that uses differential mobility classification, such as a Scanning Mobility Particle Sizer (SMPS). For a specific size channel of the spectrometer output, the penetration is the ratio of the

concentrations upstream and downstream in that channel. A continuous penetration versus particle size curve may be developed for the entire range of challenge size distribution. Huang and Chen (2001) and Agranovski et al. (2001) have shown such filtration measurement systems, which they limited to particle diameters above 30 nm.

This study deals with these two different filter test methodologies in the nano-sized particle range of 10–400 nm, including test procedure development, factors affecting variability and comparisons between results from the tests. One test system which gives a discrete penetration for a specific particle size is the TSI 8160 Automated Filter tester (updated and commercially available now as the TSI 3160) manufactured by the TSI, Inc., Shoreview, MN. Another filter test system was developed that utilizes a Scanning Mobility Particle Sizer (SMPS) to sample the particle size distributions downstream and upstream of an air filter to obtain a continuous filter penetration versus particle size curve.

Description of the instruments and methodologies

TSI 8160 Automated Filter Tester System

The TSI 8160 Automatic Filter Tester was first introduced in 1987, with the specific subject of this analysis commissioned circa 1991. It was designed for rapid and simple operation while conducting penetration versus particle diameter tests on a high efficiency air filter media. The TSI 8160 is a fully self-contained testing apparatus for conducting initial filter penetration tests of single particle size within a range between 15 and 400 nm diameter using challenge aerosols of dioctyl phthalate (DOP) and sodium chloride (NaCl). As shown in Figure 1, the TSI 8160 components are organized within two subgroups representing first, the aerosol generation group and second, the particle counting for filter penetration measurement group.

A polydisperse aerosol is created when each nozzle is singly activated from a bank of six atomizing nozzles, each nozzle drawing from a respective feedstock solution. Three of the atomizers, respectively, supply NaCl aerosol from feedstock solution concentrations of 1%, 0.1%, and 0.01% NaCl by weight in distilled water. The other three atomizers respectively supply dioctyl

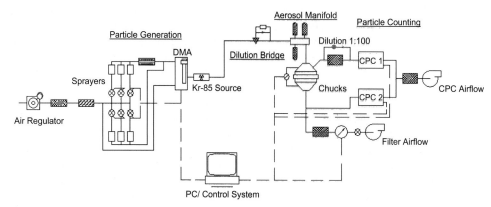

Figure 1. TSI 8160 Automated filter tester schematic.

pthalate (DOP), from individual volume concentrations of 0.3%, 0.03%, and 0.003% DOP solutions in very high purity isopropyl alcohol. Based upon the requested aerosol material and particle diameter, the computer automatically activates the associated solenoid valve for each atomizer, giving a polydisperse aerosol for subsequent size classification by a differential mobility analyzer. The NaCl aerosol flows through a diffusion dryer to ensure the formation of solid, dry particles.

For a requested computer input of type of aerosol material and particle diameter, the resulting polydisperse aerosol flows to a TSI model 3071 electrostatic classifier where a small cut is taken from the polydisperse distribution. The result is a challenge aerosol for filter testing of a selected size with a very narrow particle size distribution. Upon leaving the electrostatic classifier the aerosol is brought to a Boltzmann equilibrium charge distribution by passing through a Kr–85 charge neutralizer. Previous to the filter testing chucks, the aerosol flows through an aerosol manifold and a computer activated 100 to 1 dilution bridge to ensure the challenge aerosol is at concentration levels within the single particle counting mode capabilities of the condensation particle counters.

The TSI 8160 penetration measurement instrument group includes two condensation particle counters (TSI 3760 CPC's), pneumatic chucks, flow and pressure differential measurement, as well as vacuum pumps for flow. To increase the filter penetration range, the upstream concentration is diluted. The protocol is to use a primary split-flow diluter (apart from the dilution bridge) that is calibrated and adjusted for a 100 nm particle

diameter challenge during each set-up to give a ratio of the upstream concentration to the downstream concentration by a factor of 1:100. The computer software compensates for this dilution ratio by multiplying the upstream concentration by 100.

The percent penetration is then calculated as the ratio of downstream to upstream concentration in particles per cubic centimeter, but this only measurement is only dependable at the 100 nm particle diameter challenge where the dilution ratio was established. In reality, the performance of the CPCs may vary with respect to each other for each particle size because of particle size sensitivities and different sampling line losses. To compensate for this possible error, the manufacturer has set up a means of calculating correction factors for each requested challenge particle diameter. To generate the correction factors after the 100 nm dilution ratio is entered, the user sets up a protocol of test particle sizes and runs the filter tester without filter media in place (100% penetration). The corresponding penetration should then always be 100%, and, if it is not, a correction factor for each size is calculated as in Eq. (1) and multiplied by the measured filter percent penetration to obtain the correct penetration value.

$$CF = \frac{100}{\% \ P(\text{nofilter})} \qquad (1)$$

The test operator enters the desired particle sizes, the correction factors, the flow rate, and, after calibration of the DMA airflows, the test is ready to begin. Upon completion of a filter test battery using different discrete challenge particle sizes, the computer prints out a summary listing particle size, total concentration up and downstream, penetration, pressure drop across the media, face velocity, volumetric flow rate, and aerosol particle composition (DOP or NaCl).

TSI 8160 Calibration methodology – Requested particle diameter

In order to record the conditions of use for the TSI 8160, the initial polydisperse aerosols were measured first. A TSI model 3986 Scanning Mobility Particle Sizer (SMPS), set up for a 9.4–400 nm particle diameter range, was used to obtain aerosol size distribution for the output of each of the six atomizers. The SMPS calibration was checked with 80 nm latex sphere particles. Aerosol from each of the six atomizers was then examined with the TSI 8160 DMA classifier bypassed and is shown in Table 1.

A range of operator requested "monodisperse" particle sizes were chosen for the filter testing. The TSI 3986 SMPS was used for DOP and NaCl aerosols to give the measurement statistics for each requested size. An adequate purge time was

allowed with each new particle size to ensure all particles were flushed out of the SMPS from the previous sizing cycle. Table 2 shows the range of values obtained for geometric mean and geometric standard deviation throughout the testing period as well as an average aerosol concentration (sometimes automatically diluted by the TSI 8160). The TSI 8160 uses a TSI 3071 Electrostatic Classifier to produce filter challenge aerosols of known particle diameters. In order to give greater classifier output concentrations for the testing of very high efficiency filters with the TSI 8160, the manufacturer dictates a 2:1 ratio of the classifier sheath air to the output aerosol flow is to be used, which is very different from the 10:1 ratio specified when such classifiers are used for differential mobility particle size analysis.

In Table 2, it may be seen that the SMPS measured geometric mean diameters were sometimes different than the operator requested particle sizes. This shows the need for proper calibration for each reported particle diameter when measuring percent penetration. For the smallest requested size of both DOP and NaCl, 15 nm, the measured values were at least 25% larger, and this is the

Table 1. TSI 8160 atomizers particle size output

Particle Material	Atomizer	Program Atomizer Size Range (nm)	Percent Solution Concentration	GMD (nm)	GSD	Aerosol Concentration (#/cc)
NaCl	A	< 50	0.01	38.4	1.59	4594
NaCl	B	50–100	0.1	49.2	1.68	6987
NaCl	C	> 50	1	65.2	1.70	7710
DOP	D	< 50	0.003	36.3	1.68	4700
DOP	E	50–100	0.03	46.6	1.79	8570
DOP	F	> 50	0.3	65.4	1.77	11250

Table 2. DMA classified challenge size distributions

Operator requested particle diameter (nm)	NaCl Particles			DOP Particles		
	Actual GMD (nm)	GSD	Aerosol Concentration (#/cc)	Actual GMD (nm)	GSD	Aerosol Concentration (#/cc)
15	20.1–20.4	1.25–1.30	2040	23.9–24.8	1.54–1.58	2370
27	28.3–9.5	1.24–1.46	1900	26.1–27.6	1.37–1.47	23000
47	45.8–45.9	1.20–1.35	22600	35.9–40.0	1.50–1.65	31100
84	82.5–84.8	1.28–1.31	66400	68.7–81.6	1.31–1.60	1300
150	138.0–140.8	1.34–1.37	72000	133.0–138.0	1.37–1.40	1507
267	198.0–205.0	1.30–1.37	11700	184.0–203.0	1.36–1.61	21800

98

diameter that should be reported as the challenge particle size. The mid-range sizes, 27–84 nm, yielded better results, giving measured values within 10% for NaCl and 15% for DOP. At the larger size ranges, 150 and 267 nm, all distributions developed bimodal characteristics, and this may be due to dual charges accumulating on the large particles. This common problem cited by Yeh (1993) is caused by the larger particles' increased ability to possess multiple units of charge.

TSI 8160 Calibration Methodology – Condensation Particle Counters

Because penetration is a function of the two concentrations read by the upstream and downstream TSI 3760 CPCs, uncertainty introduced by the CPCs can have major consequences on the resulting penetration values. To determine the time required to give a stable output after being powered up (warm-up time), an experiment was designed using a Collison atomizer to produce a polydisperse 20 nm NaCl aerosol at a concentration level within the single particle counting mode range of the two CPCs. Each CPC was turned on at the same time, and 30 s concentration output intervals were recorded for each CPC and were compared to the sampling output of a warmed-up,

calibrated TSI 3010 CPC. It was found that a warm-up time of about 30–40 min was necessary to reach a stable output for the CPCs, much greater than the time it took for the TSI 3760 indicator lights to turn from red to green (indicating a stable internal temperature differential for counting).

As a check for CPC stability over time after the prescribed warm-up period, correction factors were calculated successively eight times for a range of operator requested-particle diameters, checking the stability of the counters over roughly a 1.5 h period. The results are shown in Figure 2 with the correction factor magnitude plotted versus trial number. The resulting plot shows a small amount of variability, but no more than 15% change, likely due to random error. Each day, an initial set of correction factors was calculated from three to five trial runs. As part of the test protocol, correction factors were checked periodically during the day to ensure that they were within three standard deviations of the initial daily value. Correction factor drift occurred most in particle sizes < 40 nm. The TSI 8160 factory-programmed default particle size selections use a lowest filter challenge particle size selection of 30 nm. This is a prudent choice for the lower particle size, below which a careful watch on correction factor variability is necessary.

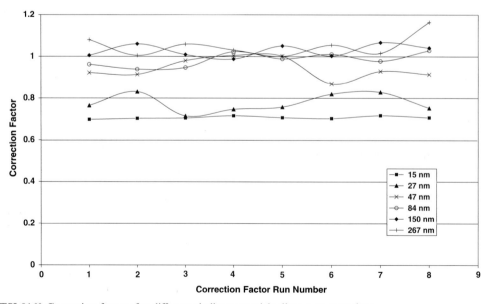

Figure 2. TSI 8160 Correction factors for different challenge particle diameters over time.

Scanning mobility particle sizer (SMPS) filtration system

A filtration penetration measurement system was constructed using an SMPS to sample downstream and upstream of a filter, the output of which was used to calculate the particle size distributions for the calculation of the subsequent penetration versus particle diameter curve for the filter.

The TSI 3936 Scanning Mobility Particle Sizer (SMPS) system used in this study consists of a number of TSI, Inc. components: 3080 Electrostatic Classifier, 3010 Condensation Particle Counter, 3080 DMA and a 0.0457 cm diameter orifice aerosol inlet impactor. During aerosol sampling, the output of the SMPS system to the TSI Aerosol Instrument Manager software consists of raw count scores from the CPC, stored in channels representing progressively greater particle sizes. These raw scores were then used by the software to generate a particle concentration versus particle diameter. The upstream filter challenge for each channel must be of a high concentration ($>10^5$ particles/cm^3) to minimize error when determining filter penetration below 0.01% penetration. If care is not taken, the subsequent calculation for the count concentration per particle size channel may vary remarkably in the lower particle size ranges for the SMPS configuration chosen, especially in the situations when the raw score counts are very low, when the counts between channels vary or go to zero and when there is not a progressive rise in counts as the particle size is increased.

As with the TSI 8160 system, the system, schematically shown in Figure 3, has an aerosol generation sub-system and filter penetration analysis sub-system. The aerosol generation system consists of two nebulizers to produce the aerosol, a small heating tube to aid in evaporation, a filtered dilution system, a Kr–85 neutralizer and a final diffusion drier to maintain the aerosol at less than 30% RH. For this study of filter penetration of nano-sized particles below 100 nm, the challenge aerosol was produced by evaporating the 70 lpm output from two BGI Inc. Collison nebulizers set at 23 psi air pressure. These were a CN60-24 jet nebulizer, (spraying lab stock distilled water, producing dissolved-solids residual particles less that 20 nm) and a CN24-3 jet nebulizer (spraying an NaCl 0.03% solution in distilled water). The CN60-24 nebulizer nozzle head was fitted with an impactor cylinder, which when mounted over the nozzles gave a nozzle-impactor distance of 1.5 mm, effectively limiting the output spray droplets to very small particles. This combination of two size distributions challenged the SMPS with neutralized aerosol number concentrations greater than 10^5 particles/cm^3 in the range of 10–172 nm. The concentration reached a 10^6 particles/cm^3 peak at 35 nm, and then gradually fell to less than 10^5 particles/cm^3 for particles above 200 nm. Collison nebulizers are well-known for their output stability and reproducibility when the temperature of the liquid solution has reached steady-state, and tests could be run for over 2 h without significant concentration changes.

The filtration penetration testing sub-system consists of pneumatic chucks to hold the filter medium with a cross-sectional area of 100 cm^2, a differential manometer and the SMPS sampling system for the sequential measurement of the particle concentration upstream and downstream of the filter sample in the chucks.

SMPS system configuration and variability

For the filter testing in this study, procedures were developed to use the SMPS output to generate a continuous curve of percent penetration versus particle diameter, essentially taking the tabulated outputs of concentration for each particle size and calculating the ratio of downstream to upstream particle concentrations for each distribution size bin. For this configuration of the TSI 3936 SMPS, the channel resolution was set to 32 channels per decade from a possible 64 in order to average and reduce noise between adjacent channels. This gave a range of particle bin midpoints of 10.4–407 nm, and a curve with 52 data points. With this configuration, each SMPS size distribution measurement scan took 135 s.

Although particle counting is an integral part of the TSI 3936 SMPS system to generate a particle size distribution, the SMPS is not an instantaneous particle counter. Correct sampling procedures must be followed to get good results, especially in particles sizes smaller than 30 nm. The SMPS sampled aerosol volume takes time to travel through the inlet, the impactor, the neutralizer, the DMA and finally through the condensation particle counter. For multiple samplings of the same

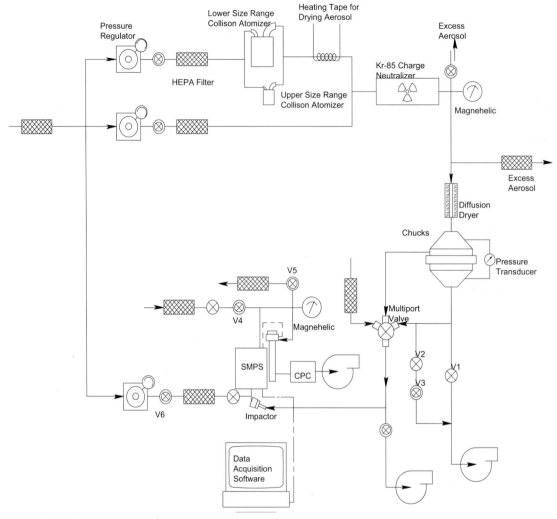

Figure 3. SMPS Filter test schematic.

aerosol, if there is not a "purge-time" pause or time interval between samples, the large classified particles left over in the system from the last sample may be erroneously counted as the smallest particles in the next sample. In addition to this serious source of inaccuracy, an even greater error may occur for the sequential sampling of different aerosol distributions, as in filtration penetration studies, where the entire aerosol sample volume of the one distribution must pass through the SMPS to then be replaced by the entire aerosol sample volume of the next distribution. This error is compounded and especially serious for filter penetration if a downstream sample is taken quickly

after an upstream sample. In that case, the large particles from the upstream sample are counted as small particles which may not exist at all in the penetrating downstream sample distribution, giving penetration results as much as two decades greater than reality.

When sampling two very different size distributions with the SMPS, the average time for the normal sample flow between samples to clean out or "purge" the TSI 3936 SMPS interior volume was found to be as long as ten minutes to minimize filter testing error. However, if when a proper clean air purging apparatus and procedure was developed (see appendix for 16 lpm filtered air purging procedure), the system could be made ready for the next distribution measurement in less than 90 s.

After a proper purging procedure is in place, other variables that may affect sampling for the filter penetration calculation must be minimized, including differences in particle loss in the sampling trains. As in the TSI 8160, a correction factor (Eq. (1))for the penetration was generated for each particle diameter channel or bin to compensate for sampling train particle loss and for possible SMPS software particle distribution calculation variability due to factors such as a loss in resolution at the upper and lower limits of the measured particle size ranges or low raw score counts at the upper or lower limits of the challenge particle size distribution.

For this SMPS filtration system, Figure 4 is a typical plot of average correction factors versus particle diameter with 95% confidence limits (dashed line curves). Below a 30 nm particle diameter, the average (solid line) correction factor increases, probably because of greater line loss differences between upstream and downstream sampling. The variability of the correction factor (dashed line) also increases greatly as the lower range of the SMPS/DMA classification system is reached. Another increase in correction factor

variability is seen above 200 nm, where, for this test, the raw score counts are rapidly decreasing at the upper end of the challenge particle sizes, causing erratic calculations of this part of the size distribution.

Results and discussion – nano-size particle air filter testing

Filter media description

Four medium-efficiency fiberglass filter papers of very low variability were used for the filter testing to give comparisons of filter penetration versus particle diameter. The media was previously donated by Hollingsworth and Voss Co. for filtration standard variability studies by the American Society for Testing and Materials (ASTM) F–21 Committee on Filtration, ASTM F1215-89 "Standard Method for Determining the Initial Efficiency of a Flatsheet Filter Medium in an Airflow Using Latex Spheres". The filter media characteristics are shown in Table 3 from studies by Japuntich (1991). The low coefficients of variation shown for both pressure drop and Q127 DOP Test penetration show the high degree of homogeneity of these samples. All filter tests were

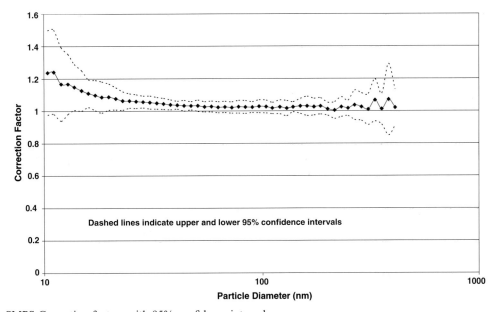

Figure 4. SMPS Correction factors with 95% confidence intervals.

Table 3. Hollingsworth and Vose filter media characteristics

Filtration parameters		H&V filter paper identification			
		HE1073	HE1021	HF0031	HF0012
Thickness (cm)	Ave.	0.053	0.069	0.074	0.074
	%COV	2.3	4.3	2.3	2.3
Basis weight (g/sq.m)	Ave.	63.9	80.3	82.6	69.2
	%COV	0.53	0.67	0.86	0.92
Pressure Drop (mmH2O) at 5.3 cm/s	Ave.	8.4	4.7	3.5	1.3
	%COV	1.48	1.35	1.94	1.47
0.3 micrometer Q127 DOP %Penetration at 5.3 cm/s	Ave.	12.8	39	45.8	79.9
	%COV	2.2	1.7	0.92	1.24
Fiber density (g/cu.cm)		2.4	2.4	2.4	2.4
Solidity		0.05	0.049	0.047	0.039
Effective fiber diameter (micrometers) (Rubow, 1981)		1.9	2.9	3.3	4.9
Effective fiber diameter (micrometers) (Davies, 1952)		2.1	3.2	3.7	5.4
Effective pore diameter (micrometers) (Benarie, 1969)		8.8	13.4	16.1	26.2

for initial penetration and were conducted with new, clean media each time with the felt side toward the challenge. Each filter sample was placed in a portable filter holder with a face area of 100 cm² and then loaded into the filter test chucks. All data were collected at volume air flows of 32 lpm, giving a face velocity of 5.3 cm/sec that is a standard for testing flat-sheet filter media.

Penetration tests: TSI 8160 NaCl challenge aerosol

The filter media were tested on the TSI 8160 for NaCl particle penetration at different, discrete particle diameters, as shown in Figure 5. Each data point is the average of ten filter tests, conducted over the course of three separate days. The challenge particle size distributions were measured at the beginning of each day of testing for each operator-requested diameter, representing three geometric mean diameter sets for each of the seven challenge sizes. The smallest particle diameter that could be made was 17.9 nm. Although some particle sizes were not as accurate with respect to the operator-requested size from the computer program, they were very precise and easily reproduced with minimal error. In Figure 5, error bars shown in the x-direction are the 95% confidence intervals for the geometric mean diameter of the particle, and the error bars in the y-direction represent the 95% confidence intervals for penetration values (error calculated according to the guidelines set forth by ASHRAE standard 52.2-1999). Less than five percent error can be expected

in generating NaCl particles on repeated tests, with 95% confidence in this estimation. The error estimated for the penetration measurements at the 95% confidence level was, in most cases, not more than 5%. However, in the case of one of the smaller particle sizes, the error climbed to nearly 8%.

A medium efficiency penetration filter media was chosen for this study because it gives penetration results over a wide span of particle diameters. The difference between the filter penetrations at 20 nm and 150 nm may be as much as two to three logarithmic decades. It would be difficult to get both penetrations on one plot if a high efficiency filter medium was used.

Penetration tests: TSI 8160 DOP challenge aerosol

Using the same procedures as shown above for NaCl, the DOP aerosol testing was also executed over the course of three days, and the actual aerosol challenge distributions of each particle size were recorded at the start of each day. The operator-requested particle sizes were the same as for NaCl. The fiberglass filter media average penetration data for the DOP aerosols looks much the same as the data for NaCl and is shown in Figure 6. However, the reproducibility of the DOP particle diameters was not as good. This may be a particular problem with our TSI 8160, but troubleshooting of the system did not improve the results.

Looking closely at Figure 6, the DOP particle size error is most prevalent at the two mid-range program sizes of 47 and 84 nm, and it is not

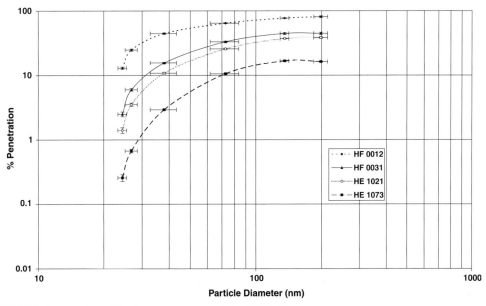

Figure 5. TSI 8160 Penetration of NaCl challenge aerosol for H&V filter media.

known why this should be so. These two particle sizes originate from different nozzles, and each nozzle is also used in at least one other particle size, which show smaller magnitudes of error. The particle size error for the other four diameters is much less, with values ranging from 4 to 7% error.

Penetration tests: SMPS system penetration test with comparisons to TSI 8160 Test

Three of the filter materials were chosen to yield a wide range of values for comparisons of percent penetration versus particle size between the SMPS

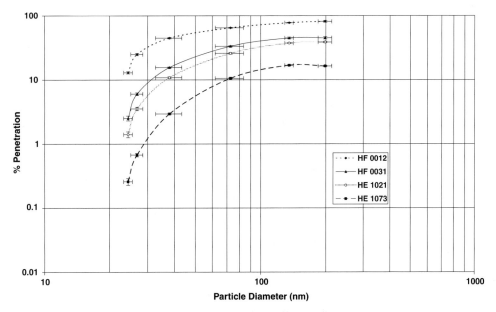

Figure 6. TSI 8160 Penetration of DOP challenge aerosol for H&V filter media.

system and the TSI 8160 system using NaCl aerosols. These were HE1073, HE1021 and HF0031. Ten samples of each filter medium were clamped in portable filter holders and were then alternately mounted into the chucks of each system, tested at face velocities of 5.3 cm/s. This yielded the plot of average values shown in Figure 7.

In Figure 7, the TSI 8160 data are shown as single points, and, since the test is actually close to a monodisperse initial penetration test, it is a good calibration tool with which to judge the validity of the SMPS Test data. The SMPS System average test results are shown as continuous curves, and the 95% confidence limits for each curve are seen as dashed-line curves. It can be seen that the average TSI 8160 data and the average SMPS System data give nearly identical results over a wide range.

It is also interesting to note in Figure 7 that the SMPS System 95 percent confidence limits increasingly diverge from the average data as the particle diameter decreases below 15–20 nm, depending on the filter medium. This increase in uncertainty is evident even though correction factors were used for the SMPS data, and it also shows that for these ranges the number of samples

should be increased to give a more confident average curve. The increase in uncertainty in Figure 7 may be the result of measurements approaching the lower limit of the DMA in the SMPS to resolve particle diameter. If another differential mobility analyzer with a suitable particle size range covering these lower diameters were used, such as a nano-DMA, it is surmised that it would give higher data confidence, but this is future work.

Figure 8 displays what happens to the SMPS system results for the HE1073 filter medium if the proper procedures of system purging and application correction factors are not followed. There is certainly more variability in the higher particle diameter range, and the results for average penetration in the lower ranges of particle diameter may be as much as 100 times greater than the uncorrected results. This is a very serious issue for SMPS and other DMA-based systems, and proper use of these systems must be followed.

Conclusions

With adequate calibration, the TSI 8160 Automated Filter Tester produces repeatable and reliable

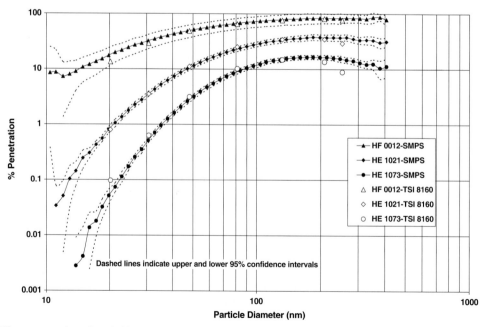

Figure 7. Filter penetration: SMPS filter test comparison to TSI 8160 for H&V filter media.

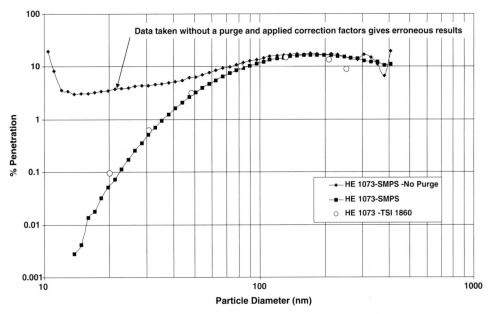

Figure 8. SMPS filter test penetration with and without purge and correction factors.

data for single particle size penetration measurement with only minor limitations. However, these limitations should be noted and kept in mind during use of the apparatus. Proper calibration of the apparatus is likely the most important step required in ensuring reliable results, as well as continual maintenance cleaning of the aerosol transport lines and the DMA. For our study, the challenge particle sizes were not necessarily the same as the program size entered by the operator. Yet, the challenge particle geometric mean diameters measured were very consistent from test to test. The user need only make particle size measurements at reasonable intervals, and the interval found to be adequate for the testing conducted in this paper was once per day.

In selecting the instrumentation for a system like the TSI 8160, the minimum detectable particle size of the condensation particle counters should be carefully considered to remove a significant degree of uncertainty from the measurements. It is recommended that the condensation particle counters be chosen with very low diameter detection limits to reduce uncertainty in correction factor estimations. Allowing a warm up period around 45 min beyond the manufacturer's pre-programmed time for the CPC's is necessary to obtain a stable response comparison between the two counters.

For the TSI 8160 Test, the variability of the particle size repeatability was greater for DOP than for NaCl aerosols, which may be due to some peculiarity in our test. Although the theoretical prediction of the output of a DMA classification system may be used to create an aerosol in a certain size range, it is always prudent to measure the particle diameter and to implement proper calibration procedures to ensure good size data. The TSI 8160 technique of obtaining a higher concentration output by using a 2:1 ratio of sheath air to monodisperse aerosol output in the TSI 3071 differential mobility analyzer classifier may give an uncertainty in the size distribution of the "monodisperse" output aerosol.

A filter penetration measurement system has been assembled using a Scanning Mobility Particle Sizer (SMPS), which for a filter medium yields a continuous percent penetration versus particle diameter curve from the sampling of upstream and downstream particle size distributions. A procedure (see appendix) has been implemented to minimize error by using proper purging between particle samples and employing a method of applying correction factors to penetration for different particle sizes. Results of percent penetration versus particle size from the testing of flat sheet filters have shown excellent comparison with TSI 8160 data.

For both of these nano-sized particle air filtration tests, the penetration versus particle size curves for the filter media studied follow the trends of the theoretical Brownian capture model of decreasing penetration with decreasing particle diameter below 100 nm. There was no evidence of "particle bounce". However, if the SMPS system procedures implemented are disregarded, erroneous results appearing as "particle bounce" should be expected, with values of this subsequently invalid filter penetration ranking much higher than reality in ranges below 30 nm.

Acknowledgements

This research was done as part of a Center for Filtration Research liaison between the University of Minnesota and 3M Company. We would also like to thank Tom Kotz of Donaldson Company, Inc., Minneapolis, MN, for his thoughtful suggestions on the need to purge the electrostatic neutralizer on scanning mobility analyzers.

Appendix: SMPS System filter test procedure

The SMPS system is used to generate distributions of concentration versus particle size channel by aerosol sampling downstream and upstream of an air filter medium. The filter penetration is the ratio of the downstream to upstream particle concentration for each size channel. The SMPS system test procedure is outlined below with references to the schematic, Figure 3.

1. System warm up

- Observe the SMPS and CPC warm up period as outlined by manufacturer.
- Run Collison atomizers 10 min to allow liquid reservoir temperatures to reach a steady-state.

2. Filter air flow

- Set the prescribed filter test flow using valve V1, with the multiport valve set for sampling from the downstream position.
- To maintain the filter test flow during the sequential upstream and downstream change

sampling, a downstream vacuum flow is turned on using valve V3 during upstream sampling and turned off during downstream sampling, with this flow adjusted using valve V2.

3. SMPS Internal Volume System Purge

- An SMPS system purge is required each time the multiport valve is switched to another sampling location (upstream or downstream).
- After setting the multiport valve to a sampling location, bleed valve V4 is opened to bring the SMPS internal system air pressure slowly up to atmospheric pressure (if a sampling inlet impactor is used).
- Valve V5 is opened to allow for excess purge flow, and bleed valve V4 is closed.
- Clean air flow Valve V6 is then turned on to purge the SMPS with 16 lpm filtered air for 45 s.
- Clean air valve V6 is closed, valve V5 is closed and the system is given another 45 seconds for the internal pressure to equilibrate to the internal negative pressure (if a sampling inlet impactor is used) and when no counts are read on the CPC.

4. Upstream and downstream concentration measurements

- The filter medium is placed in the chucks.
- The multiport valve is switched to the downstream location, valve V3 is closed.
- A system purge is done.
- Two SMPS data sets are taken from the downstream sampling location.
- The multiport valve is switched to the upstream location, valve V3 is open.
- A system purge is done.
- Two SMPS data sets are taken from the upstream sampling location.
- The percent penetration is calculated for each size channel as 100 times the ratio of the data set averages of the downstream to upstream concentrations.

5. Percent penetration correction for downstream and upstream sampling differences

- To compensate for particle loss differences in sampling locations, correction factors are calculated for each SMPS particle size channel from a penetration test with no media present, a theoretical 100 percent penetration, by the following equation:

$$CF = \frac{100}{\%\ P(\text{nofilter})} \quad (1)$$

- Averages of five correction factor data sets taken.
- The penetrations for each particle size channel are then multiplied by the respective average correction factors to give the final corrected penetrations.

References

Agranovski, I.E., R.D. Braddock & T. Myojo, 2001. Comparative study of the performance of nine filters utilized in filtration of aerosol by bubbling. Aerosol Sci. Technol, 35, 852–859.

AHSRAE, 1999. Method of Testing General Ventilation Air-Cleaning Devices for Removal. Efficiency by Particle Size. (ASHRAE Standard 52.2). ASHRAE Inc., Atlanta, GA.

Aitken R.J., K.S. Creely & C.L. Tran, 2004. Nanoparticles: An occupational hygiene review. HSE Report 274. Health and Safety Executive, UK.

ASTM., 2001. Standard Methods or Gas Flow Resistance Testing of Filtration Media. (ASTM Standard F778-88). ASTM International, West Conshohocken, PA.

ASTM., 2005. Standard Test Method for Grease Particle Capture Efficiency of Commercial Kitchen Filters and Extractors, (ASTM F2519–05). ASTM International, West Conshohocken, PA.

Benarie, M. & Staub R. Luft, Influence of Pore Struture Upon Separation Efficiency in Fiber Filters. 1969. 29 (2), p. 37.

Brown, R.C., 1993. Aerosol Filtration, Chapter 4, Pergamon, Oxford.

Cheng, Y.S. & H.C. Yeh, 1980. Theory of a screen-type diffusion battery. J. Aerosol Sci. 11, 313.

Davies, C.N., 1952. Proc. Inst. Mech. Engrs., London, 1B, p. 185.

Dhaniyala, S. & B.Y.H. Lui, 1999. Investigations of particle penetration in fibrous filters. J. of IEST 42(1), 32–40.

Franklin, L., 2005. Characterization and Optimization of a TSI Model 8160 Automated Filter Tester for Nanoparticle Filtration Studies. MS Plan B Paper, Mechanical Engineering, University of Minnesota, Minneapolis, MN.

Heim, M., B.J. Mullins, M. Wild, J. Meyer & G. Kasper, 2005. Filtration efficiency of aerosol particles below 20 nanometers. Aerosol Sci. Technol. 39, 782–789.

Hinds, W.C., 1999. Aerosol Technology: Properties, Behavior, and Measurement of Airborne Particles, Chapter 9 (2nd ed.). John Wiley and Sons, New York.

Huang, S.H. & C-C Chen, 2001. Filtration characteristics of a miniature electrostatic precipitator. Aerosol Sci. Technol. 35, 792–804.

Japuntich, D.A., 1991. Particle Clogging of Fibrous Filters. Ph.D Thesis. Loughborough, Loughborough University of Technology, U.K.

Kim, S.C., M.S. Harrington, A. Rengasamy & D.Y.H. Pui, 2005. Collection Efficiency of Filter Media for Nanoscale Particles. Second International Symposium on Nanotechnology and Occupational Health. Minneapolis, MN.

Kirsch, V.A. & I.B. Stechkina, 1978. In: Shaw D. T. ed. Fundamentals of Aerosol Science, Chapter 4. Wiley, New York, p. 165.

Kirsh, V.A., 2003. Deposition of Aerosol Nanoparticles in Fibrous Filters. Colloid Journal (Kolloidnyl Zhurnal) 65(6), 726–732.

Lee, K.W. & B.Y.H Liu, 1980. J. Air Poll. Control Assoc. 30, 377–381.

Lifshutz, N. & M. Pierce, 1996. A general correlation of MMPS penetration as a function of face velocity with the Model 8140 using the CertiTest 8160. 24th DOE/NRC Nuclear Air Cleaning and Treatment Conference, Portland, OR, 698–707.

Rubow, K., 1981. Submicron Aerosol Filtration Characteristics of Membrane Filters, Ph.D. Thesis, U. of MN, pp. 37–38.

Skaptsov, A.S., A.M. Baklanov, S.N. Dubtsov, N. S. Laulainen, G. Sem & S. Kaufman, 1996. An experimental study of the thermal rebound effect of nanometer aerosol particles. J. Aerosol Sci. 27(S1), S145–S146.

Wang, H.C., 1996. Comparison of thermal rebound theory with penetration measurements of nanometer particles through wire screens. Aerosol Sci. Technol. 24, 129–134.

Wang, H.C. & G. Kasper, 1991. Filtration Efficiency of Nanometer-Size Aerosol Particles. J. Aerosol Sci. 22, 31–41.

Yeh, H.C., 1993. Electrical Techniques. In: Willeke K., & Baron E.A. eds. Aerosol Measurement: Principles, Techniques, and Application. Van Nostrand Reinhold, New York, NY, pp. 410–426.

Journal of Nanoparticle Research (2007) 9:109–115
DOI 10.1007/s11051-006-9155-9

Special focus: Nanoparticles and Occupational Health

Modeling of filtration efficiency of nanoparticles in standard filter media

J. Wang[1,*], D.R. Chen[2] and D.Y.H. Pui[1]

[1]*Department of Mechanical Engineering, University of Minnesota, Minneapolis, MN, USA;* [2]*Department of Mechanical Engineering, Washington University at St. Louis, St. Louis, MO, USA;* *Author for correspondence (E-mail: wangj@aem.umn.edu)*

Received 26 July 2006; accepted in revised form 2 August 2006

Key words: filtration efficiency modeling, nanoparticle penetration, standard filter media, occupational health

Abstract

The goal of this study is to model the data from the experiments of nanoparticle filtration performed at the Particle Technology Lab, University of Minnesota and at the 3M Company. Comparison shows that the experimental data for filter efficiency are bounded by the values computed from theoretical expressions which do not consider thermal rebound. Therefore thermal rebound in the tested filter media is not detected down to 3 nm particles in the present analysis. The efficiency measured experimentally is in good agreement with the theoretical expression by Stechkina (1966, Dokl. Acad. Nauk SSSR 167, 1327) when the Pectlet number *Pe* is larger than 100; it agrees well with the theoretical expression by Kirsch and Stechkina (1978, Fundamentals of Aerosol Science. Wiley, New York) when *Pe* is of the order of unit. We develop an empirical power law model for the efficiency depending on the Peclet number, which leads to satisfactory agreement with experimental results.

Introduction

Aerosol filtration is used in diverse applications, such as respiratory protection, air cleaning of smelter effluent, processing of hazardous material, and clean rooms. The filtration of nanoparticles is becoming an important issue as they are produced in large numbers from material synthesis and combustion emission. They may pose a health risk because nanoparticles can readily enter the human body through inhalation and their toxicity is relatively high due to the large specific surface area (Oberdörster et al., 2005; Maynard & Kuempel, 2005).

Some studies show that nanoparticles may bounce through the filter media due to their high thermal speed. The theoretical study of Wang and Kasper (1991) predicted the onset of aerosol rebound below 10 nm during filtration process. Balazy et al. (2004) reported that thermal bounce occurred for 20 nm particles. There are also studies showing that the thermal bounce is not detected down to 2–3 nm particles. Ichitsubo et al. (1996) carried out experiments using wire screens and did not detect measurable thermal bounce until 3 nm and lower. Alonso et al. (1997) used a tandem DMA technique, and detected no particle bounce in the same size range of Ichitsubo et al. (1996). Heim et al. (2005) examined the filtration efficiency of sub-20 nm aerosol particles on grounded metal filters and screens, and plastic mesh. They found that even for particle sizes as small as 2.5 nm there was no measurable deviation from the classical single-fiber-efficiency theory, thus no measurable thermal bounce.

Experiments have been performed at the University of Minnesota (UMN) and the 3M Company (see the companion papers by Kim et al. 2006 and Japuntich et al. 2006) to investigate the filtration efficiency of standard filters. The size of the aerosol particles is 3–20 nm in the UMN study and is 15–400 nm in the 3M study. We compare the experimental data with theoretical filtration models. Since diffusion is the dominant deposition mechanism for nanoparticle filtration, the expressions for efficiency due to diffusion play an essential role in the comparison. Many researchers obtained theoretical expressions for the efficiency due to diffusion by solving the convective diffusion equation using a boundary layer approach. Friedlander (1957, 1958) and Natanson (1957) performed such analyses using the Stokes flow past a single fiber. Fuchs and Stechkina (1963), Pich (1965), Stechkina (1966), Kirsch and Stechkina (1978) and Lee and Liu (1982) used the Kuwabara flow field, which allows the consideration of the interference effect by the neighboring fibers. The comparison shows that the experimental data for filter efficiency are bounded by the values computed from theoretical expressions.

Experimental results

Four standard filter media, HE1073, HE1021, HF0031 and HF0012 manufactured by Hollingsworth and Vose Fiberglass Media are used in the experiments. HE-type filter media approach HEPA regime for small aerosol particles; HF-type filter media are more common to standard HVAC systems. The filtration efficiency of these filter media has been tested using ASTM test method F1215-89. The results showed low coefficients of variation (COV) and high degree of homogeneity. Characteristic parameters for these filter media are given in Table 1. The solidity is the volume fraction of fibers, which is equal to (1 – porosity). The standard filter media are composed of fibers of different sizes with random orientations. The effective fiber diameter used in our calculation was computed based on the pressure drop and includes the effects of inhomogeneity.

Filtration testing systems have been set up at UMN and the 3M Company. Silver particles in the range of 3–20 nm were used in the UMN study and NaCl particles in the range of 15–400 nm were used in the 3M study. The details of the experimental setup are documented in the companion papers by Kim et al. (2006) and Japuntich et al. (2006).

Measured filter penetration is plotted as a function of the aerosol particle size in Figure 1a and b for face velocities 5.3 cm/s and 10 cm/s, respectively. The data from UMN and 3M for particle size ∼20 nm agree very well when the face velocity is 10 cm/s; the agreement is less good but still satisfactory when the face velocity is 5.3 cm/s. The values of the Reynolds number at different face velocities are listed in Table 1. They are much smaller than 1, indicating that the flows are in Stokes flow regime and the air inertia is negligible.

Table 1. Characteristic parameters of the four filter media and the Reynolds number at different face velocities

Parameters		HE1073	HE1021	HF0031	HF0012
Thickness (cm)	Average	0.053	0.069	0.074	0.074
	% COV	2.3	4.3	2.3	2.3
Solidity	–	0.05	0.049	0.047	0.039
Effective fiber diameter (μm)	–	1.9	2.9	3.3	4.9
DOP % Penetration 0.3 μm at 5.3 cm/s	Average	12.8	39	45.8	79.9
	% COV	2.2	1.7	0.92	1.24
Pressure Drop at 5.3 cm/s (mmH$_2$O)	Average	8.4	4.7	3.5	1.3
	% COV	1.48	1.35	1.94	1.47
Fiber Density (g/cm^3)	–	2.4	2.4	2.4	2.4
Reynolds number at different face velocities	5.3 cm/s	0.0068	0.010	0.012	0.018
	10 cm/s	0.013	0.020	0.022	0.033
	15 cm/s	0.019	0.029	0.034	0.050

The Reynolds number is defined as $R_e = U_0 d_f / v$, where U_0 is the face velocity, d_f is the fiber diameter, and v is the kinematic viscosity of air

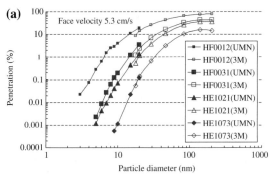 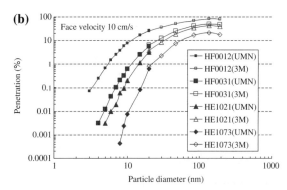

Figure 1. The penetration as a function of the aerosol particle size for the four standard filter media: HE1073, HE1021, HF0031 and HF0012. The face velocity is 5.3 cm/s in (a) and 10 cm/s in (b). The size of the aerosol particles is 3–20 nm in UMN experiments (solid symbols in the figures) and is 20–260 nm in 3M experiments (open symbols in the figures).

The filter penetration is related to the single fiber efficiency through

$$P = \exp\left(-\frac{4\alpha Et}{\pi d_f}\right), \quad (1)$$

where P is the penetration, α is the solidity, E is the single fiber efficiency, t is the thickness of the filter and d_f is the fiber diameter. Inverting Eq. (1), we can express the single fiber efficiency as

$$E = -\frac{\pi d_f}{4\alpha t} \ln P. \quad (2)$$

Using (2), we compute the single-fiber efficiency E from the UMN and 3M experimental data.

Modeling of the single fiber efficiency for nanoparticles

The total single-fiber efficiency E has contributions from different collection mechanisms and can be written as

$$E \approx E_D + E_R + E_{DR} + E_I + E_G, \quad (3)$$

where E_D, E_R, E_I, E_G represent the collection efficiency due to diffusion, interception, inertial impaction and gravity, respectively; E_{DR} accounts for the enhanced collection due to interception of the diffusing particles. Expressions for these efficiencies can be found in Hinds (1998) and are used to estimate the relative importance of different collection mechanisms. For example, if we consider the filter media HF0012 at the face velocity of 10 cm/s, E_D accounts for 99.85% of E when the

particle diameter is $d_p = 1$ nm, 85.5% of E when the particle diameter is $d_p = 100$ nm.

Since diffusion is the dominant filtration mechanism for nanoparticles, the total efficiency E computed using (2) is approximately equal to E_D when $d_p < 100$ nm. In the literature, a number of theoretical expressions for the single fiber efficiency due to diffusion are derived. Hinds (1998) gives the following expression

$$E_D = 2Pe^{-2/3}, \quad (4)$$

where Pe is the Peclet number and defined as

$$Pe = \frac{d_f U_0}{D} = d_f U_0 \frac{3\pi\eta}{kTC_c} d_p, $$

where D is the diffusion coefficient, η is the viscosity of air, k is the Boltzmann constant, T is the temperature, and C_c is the slip correction factor. The Peclet number provides an indication of the relative importance of diffusion and convection. In our calculation, the expression for the slip correction factor obtained by Kim et al. (2005) is used. Stechkina (1966) did a boundary layer analysis for the convective diffusion equation and obtained

$$E_D = 2.9 Ku^{-1/3} Pe^{-2/3} + 0.624 Pe^{-1}, \quad (5)$$

where Ku is the Kuwabara hydrodynamic factor. Kirsch and Stechkina (1978) considered the effects of the gas slip and the case in which

$$(2k_1/Pe)^{1/3} < Kn, \quad (6)$$

where $Kn = 2\lambda/d_f$ is the Knudsen number, $k_1 = Ku + \tau Kn$ and τ is a coefficient of order of unity and depends on interaction of gas and fiber. Note that $\delta_1 = (2k_1/Pe)^{1/3}$ is the characteristic length of the diffusion boundary layer. Kirsch and Stechkina (1978) presented an analysis of the convective diffusion equation and showed that

$$E_D = 3.20k_1^{-1/2}(\tau Kn)^{1/2}Pe^{-1/2}. \quad (7)$$

We assume $\tau = 1$ in our calculations. The prominent feature of Eq. (6) is that the power of Pe is $-1/2$, not $-2/3$ as in Eqs. (4) and (5).

We compare the efficiency obtained from the UMN and 3M experiments to the theoretical expressions (4), (5) and (7). We focus on particles smaller than 100 nm, for which $E \approx E_D$. The theoretical studies suggest

$$E = \text{Function}(Pe, \alpha, Kn).$$

The efficiency E is plotted against the Peclet number Pe for the filter media HE1073, HE1021,

HF0031 and HF0012 in Figure 2a, b, c and d, respectively. The solidity α and the Knudsen number Kn for these filters are listed in each figure.

Figure 2 shows that generally Eq. (5) overestimates the efficiency and Eq. (7) underestimates it. The experimental results for E are in good agreement with Eq. (5) when $Pe > 100$; they agree well with Eq. (7) when Pe is of the order of unit. The good agreement with Eq. (5) at high Pe seems to be reasonable, since Eq. (5) was derived from a boundary layer analysis which is presumably valid at high Pe. The agreement with Eq. (7) at low Pe is better when Kn is larger. It shows that the gas slip should be considered if Kn is not negligible. We notice that the condition (6) is not satisfied at low Pe in the experiments, thus the good agreement with Eq. (7) at low Pe is not clearly understood. The experimental data for filter efficiency are bounded by the values computed from theoretical expressions (5) and (7). Therefore thermal rebound

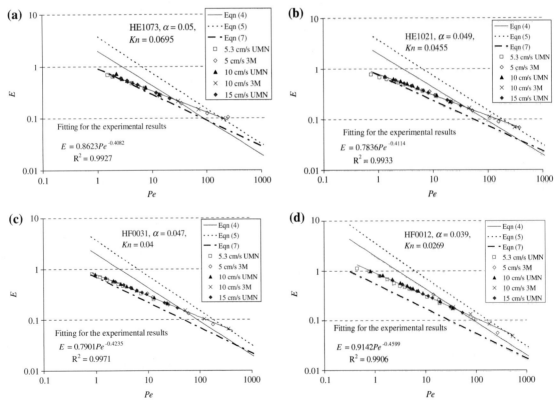

Figure 2. The efficiency E against the Peclet number Pe for the filter media (a) HE1073, (b) HE1021, (c) HF0031 and (d) HF0012. The theoretical expressions (4), (5) and (7) are represented by lines and the experimental results are indicated by symbols. The experimental results include data obtained using the face velocities 5.3, 10 and 15 cm/s.

in the tested filter media is not detected down to 3 nm particles.

We seek a new empirical model which can describe the experimental data more accurately. Figure 2 shows that the experimental results can be satisfactorily fitted using power laws

$$E = AP_e^m. \tag{8}$$

The fitting parameters A and m are listed in Table 2 for the four standard filter media.

If we take the average values of A and m, a power law in the following form can be obtained

$$E = 0.84Pe^{-0.43}. \tag{9}$$

Equation (9) can be used to compute the penetration. In Figure 3, the computed penetration is compared with the experimental results. At the same time, the equation $E = 2Pe^{-2/3}$ in Hinds (1998) is also used to compute the penetration and compared to the experimental results. The agreement for the equation $E = 2 Pe^{-2/3}$ is not good, especially at small particle sizes. The agreement for the equation $E = 0.84 Pe^{-0.43}$ is satisfactory. This comparison demonstrate the simple power law expression with only two fitting parameters can lead to good agreement with a wide range of filtration data, which cover all the four filter media and particles smaller than 100 nm at face velocities of 5.3, 10 and 15 cm/s.

Processing the data using dimensionless groups

Friedlander (1967, 2000) analyzed the convective diffusion equation and derived a functional relationship which covers the filtration efficiency due to diffusion and interception. It indicates that the group $(E R Pe)$ should be a single-valued function of $(R Pe^{1/3})$:

$$ERPe = Function(RPe^{1/3}), \tag{10}$$

over the range $Pe \gg 1$, $R \ll 1$ for a fixed Reynolds number smaller than 1. Here $R = d_p/d_f$ is the interception parameter. In the limiting case $R \to 0$, diffusion is the dominant deposition mechanism, therefore the efficiency is independent of R and (10) should have the form

$$ERPe \sim (RPe^{1/3})^1. \tag{11}$$

In the limiting case $Pe \to \infty$, interception is far more important than diffusion, therefore the efficiency is independent of Pe and Eq. (10) suggests

$$ERPe \sim (RPe^{1/3})^3. \tag{12}$$

If the group $(E R Pe)$ is plotted against $(R Pe^{1/3})$, all data at a fixed Reynolds number should fall on the same curve. On a log–log plot, the slope of the curve is unity for small values of $(R Pe^{1/3})$ and three for large values of $(R Pe^{1/3})$.

We plot the experimental results using the dimensionless groups $(E R Pe)$ and $(R Pe^{1/3})$ in Figure 4. The data points for each filter media almost fall on the same curve, in agreement in the functional relation (10). The curves for the four filter media are close, but somewhat different. This indicates that the filter solidity plays a role in determining the functional relation (10).

We notice that the slopes of the curves become larger as the value of $(R Pe^{1/3})$ increases. For convenience, we select $R Pe^{1/3} = 0.1$ as a limiting value. The slopes of the curves for the four filter media are in the range of (1.25–1.32) when $R Pe^{1/3} < 0.1$ and in the range of (1.37–1.68) when $0.1 < R Pe^{1/3} < 2$. It appears that our data for $R Pe^{1/3} < 0.1$ are close to Eq. (11). Our data for $0.1 < RPe^{1/3} < 2$ are in a transition regime between the diffusion-dominated regime and the interception-dominated regime. We do not have data for the interception-dominated regime to confirm Eq. (12).

Discussion and conclusion

In this study we process the experimental data for fibrous filtration of aerosol particles in the range of 3–260 nm. The single-fiber efficiency is computed from the experimental data and compared to well-known theoretical expressions. The experimental results are in good agreement with the theoretical

Table 2. The solidity α the Knudsen number Kn, and the fitting parameters A and m for the four standard filter media

Filter	α	Kn	A	m
HE1073	0.05	0.0695	0.8623	−0.4082
HE1021	0.049	0.0455	0.7836	−0.4114
HF0031	0.047	0.04	0.7901	−0.4235
HF0012	0.039	0.0269	0.9142	−0.4599

114

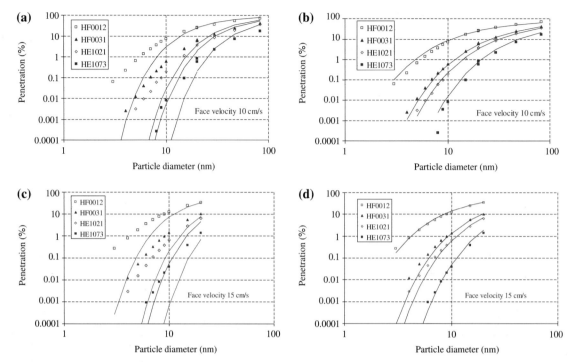

Figure 3. The penetration as a function of the particle size. The face velocity is 10 cm/s in (a) and (b) and 15 cm/s in (c) and (d). The symbols are experimental results and the solid lines are computed using different expressions for single-fiber efficiency. Hinds equation $E = 2Pe^{-2/3}$ is used in (a) and (c); the agreement is not good. The power law $E = 0.84\ Pe^{-0.43}$ is used in (b) and (d); the agreement is satisfactory.

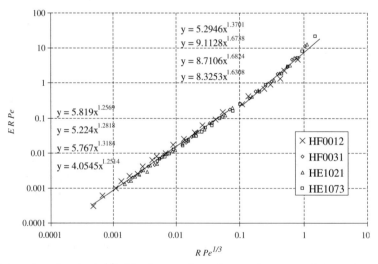

Figure 4. The group $(E\ R\ Pe)$ against $(R\ Pe^{1/3})$. The data are from the experiments at UMN and the 3M Company (see the companion papers by Kim et al. (2006) and Japuntich et al. (2006)).

expression by Stechkina (1966) when the Pectlet number $Pe > 100$; they agree well with the theoretical expression by Kirsch and Stechkina (1978) when Pe is of the order of unit. The penetration decreases as the particle size approaches 3 nm and the single-fiber efficiency is bounded by the values computed from theoretical expressions. Thus no evidence of thermal rebound is detected in the tested filter media for particles as small as 3 nm.

The inhomogeneity and random orientations of fibers are difficult to account for. Their effects are only considered in the calculation of the effective fiber diameter but not considered in the present analysis of the single-fiber efficiency. This may be part of the reason why the experimental efficiency does not agree very well with theoretical expressions.

Numerical simulation provides another approach for filtration studies. Kirsh (2003) numerically solved the convective diffusion equation in a two-dimensional cell model. His results are in good agreement with the Stechkina formula (5) even when Pe is of the order of unit. But our experimental data do not agree well with (5) in the same range of Pe. It appears that accurate modeling of filtration by standard filters at small Pe limit is still a challenging task.

We found that a power law model for the efficiency depending on the Peclet number leads to satisfactory agreement with experiments. The power of the Peclet number is in the range of -0.41 to -0.46 for particles smaller than 100 nm. The experimental data are also consistent with the functional relation suggested by Friedlander (1967, 2000).

Acknowledgement

The authors thank the support of members of the Center for Filtration Research: 3M, Donaldson, Fleetguard, Samsung Digital Appliance, Samsung Semiconductor, TSI, and W. L. Gore & Associates. We also thank Prof. Sheldon K. Friedlander for valuable discussions.

References

Alonso M., Y. Kousaka, T. Hashimoto & N. Hashimoto, 1997. Penetration of nanometer-sized aerosol particles through wire screen and laminar flow tube. Aerosol Sci. Technol. 27, 471–480.

Balazy A., A. Podgorski & L. Gradon, 2004. Filtration of nanosized aerosol particles in fibrous filters. I – experimental results. J. Aerosol Sci. EAC Proceedings Vol. II, S967–S980.

Friedlander S.K., 1957. Mass and heat transfer to single spheres and cylinders at low Reynolds numbers. AIChE J. 3, 43–48.

Friedlander S.K., 1958. Theory of aerosol filtration. Ind. Eng. Chem. 50, 1161–1164.

Friedlander S.K., 1967. Particle diffusion in low speed flows. J. Colloid Interface Sci. 23, 157–164.

Friedlander S.K., 2000. Smoke, Dust, and Haze, 2nd edn. Oxford University Press.

Fuchs N.A. & I.B. Stechkina, 1963. A note on the theory of fibrous aerosol filters. Ann. Occup. Hyg. 6, 27–30.

Heim M., B. Mullins, M. Wild, J. Meyer & G. Kasper, 2005. Filtration efficiency of aerosol particles below 20 Nanometers. Aerosol Sci. Technol. 39, 782–789.

Hinds W.C., 1998. Aerosol technology, 2nd edn. Wiley-interscience.

Ichitsubo H., T. Hashimoto, M. Alonso & Y. Kousaka, 1996. Penetration of ultrafine particles and ion clusters through wire screens. Aerosol Sci. Technol. 24, 119–127.

Japuntich D., L. Franklin, D.Y.H. Pui, T. Kuehn & S.C. Kim, 2006. Air filtration testing using the TSI 8160 automated filter tester for solid and liquid aerosols of 15 to 400 nm diameter. J. Nanoparticle Research, Special Issue: Nanotechnology and occupational health.

Kim J.H., G.W. Mulholland, S.R. Kukuck & D.Y.H. Pui, 2005. Slip correction measurement of certified PSL nanoparticles using a nanometer differential mobility analyzer (Nano-DMA) for Knudsen number from 0.5 to 83. J. Res. Natl. Inst. Stand. Technol. 110, 31–54.

Kim S.C., M. Harrington & D.Y.H. Pui, 2006. Filter collection efficiency for nanoscale particles. J. Nanoparticle Research, Special Issue: Nanotechnology and occupational health.

Kirsch A.A. & I.B. Stechkina, 1978. The theory of aerosol filtration with fibrous filters. In: Shaw D.T. ed. Fundamentals of Aerosol Science. Wiley, New York.

Kirsh V.A., 2003. Deposition of aerosol nanoparticles in fibrous filters. Colloid J. 65, 726–732Translated from Kolloidnyi Zhurnal, 65, 795–801.

Lee K.W. & B.Y.H. Liu, 1982. Theoretical study of aerosol filtration in fibrous filters. Aerosol Sci. Technol. 1, 147–161.

Maynard A.D. & E.D. Kuempel, 2005. Airborne nanostructured particles and occupational health. J. Nanoparticle Res. 7, 587–614.

Natason G.L., 1957. Diffusional precipitation of aerosols on a streamlined cylinder with a small capture coefficient. Proc. Acad. Sci. USSR, Phys. Chem. Sec. 112, 21–25Dokl. Akad. Nauk, SSSR 112:100.

Oberdörster G., E. Oberdörster & J. Oberdörster, 2005. Nanotoxicology: An emerging discipline evolving from studies of ultrafine particles. Environ. Health Perspect. 113, 823–839.

Pich J., 1965. The filtration theory of highly dispersed aerosols. Staub Reinhalt. Luft. 5, 16–23.

Stechkina I.B., 1966. Diffusion precipitation of aerosols in fiber filters. Dokl. Acad. Nauk SSSR 167, 1327.

Wang H.C. & G. Kasper, 1991. Filtration efficiency of nanometer-size aerosol particles. J. Aerosol Sci. 22, 31–41.

Journal of Nanoparticle Research (2007) 9:117–125
DOI 10.1007/s11051-006-9176-4

© Springer 2006

Special issue: Nanoparticles and Occupational Health

Experimental study of nanoparticles penetration through commercial filter media

Seong Chan Kim, Matthew S. Harrington and David Y. H. Pui*
*Particle Technology Laboratory, Department of Mechanical Engineering, University of Minnesota, 111 Church St. S.E., Minneapolis, MN, 55455, USA; *Author for correspondence (Tel.: +64-3-364-2507; E-mail: dyhpui@tc.umn.edu)*

Received 29 August 2006; accepted in revised form 1 September 2006

Key words: nanoparticles, penetration, filtration, thermal rebound, diffusion, occupational health

Abstract

In this study, nanoparticle penetration was measured with a wide range of filter media using silver nanoparticles from 3 nm to 20 nm at three different face velocities in order to define nanoparticle filtration characteristics of commercial fibrous filter media. The silver particles were generated by heating a pure silver powder source via an electric furnace with a temperature of 870°C, which was found to be the optimal temperature for generating an adequate amount of silver nanoparticles for the size range specified above. After size classification using a nano-DMA, the particle counts were measured by an Ultrafine Condensation Particle Counter (UCPC) both upstream and downstream of the test filter to determine the nanoparticle penetration for each specific particle size. Particle sampling time continued long enough to detect more than 10^5 counts at the upstream and 10 counts at the downstream sampling point so that 99.99% efficiency can be detected with the high efficiency filter. The results show a very high uniformity with small error bars for all filter media tested in this study. The particle penetration decreases continuously down to 3 nm as expected from the classical filtration theory, and together with a companion modeling paper by Wang et al. in this same issue, we found no significant evidence of nanoparticle thermal rebound down to 3 nm.

Introduction

Nanotechnology, which involves the manipulation of matter at nanometer length scales to produce new materials, structures and devices, has the potential to start the new industrial revolution. The potential for new products leading to improvements in our lives is astounding. Nanoparticles often behave much differently than bulk samples of the same materials, resulting in unique electrical, optical, chemical, and biological properties. The special properties of nanoparticles give rise to recent concerns about the potential health hazards posed to workers or users that are exposed to them (Oberdörster, 2000; Maynard, 2003). Therefore, nanoparticle research has received considerable attention in many laboratories and industrial fields, especially for studying the health effects of nanoparticles and their control.

Filtration is the simplest and most common method for air cleaning, and aerosol filtration is used in diverse applications, such as respiratory protection, air cleaning of smelter effluent, processing of nuclear and hazardous materials, and clean rooms. However, the process of filtration is complicated, and although the general principles are well known there is still a gap between theory and experiment (Hinds, 1999). In particular, recent

118

modeling and experiments pointed to the potential penetration of nanoparticles through the filters due to thermal rebound. Further, nanoparticle penetration has not been shown clearly due to the difficulties of system set-up and penetration measurement. Wang and Kasper (1991) suggested a numerical model for nanoparticle penetration showing that the thermal impact velocity of a particle will exceed the critical sticking velocity in the size range between 1 and 10 nm depending sensitively on elastic and surface adhesion parameters. Ichitsubo et al. (1996) conducted an experimental work of nanoparticle penetration using wire screens, and showed the nanoparticle penetration below two nm in size was higher than the theoretical results due to the thermal rebound. Following this, Alonso et al. (1997) used a tandem DMA technique, and detected no particle bounce effects in the same size range as Ichitsubo et al. As of now, the thermal rebound effect on nanoparticle filtration is not well proven, and it is very important to study the air filtration properties of nanoparticles to determine the filtration requirements of personal protective equipment or HVAC filter.

In this study, the nanoparticle penetration test system has been established and the nanoparticle penetration was tested with a wide range of filter media (four fiberglass filter media, four electret filter media and one nanofiber filter media) using silver nanoparticles from 3 nm to 20 nm at face velocities of 5.3, 10 and 15 cm/s.

Experiments

Figure 1 shows a schematic diagram of a nanoparticle filtration test system, and it consists of a nanoparticle generation system, a size classification system and a penetration measurement system. An electric furnace is used to generate silver nanoparticles from a pure silver powder source (99.999%, Johnson Mattney Electronics), and clean compressed air is used as a carrier gas with flow rate of 3.0 lpm. The silver powder source located in the center of a heating tube is vaporized and condensed into silver nanoparticles with a wide size distribution at stainless steel tube parts. The particle size distribution can be controlled by changing the furnace temperature as shown in Figure 2. The average size and the particle number concentration of silver nanoparticles generated by the furnace increases with the furnace temperature, because at a higher temperature, the evaporation rate increases, giving rise to a larger amount of condensable vapor which allows the particles to grow to larger sizes by agglomeration and condensation (Ku & Maynard, 2006).

The silver nanoparticles are given a Boltzman charge distribution by Po-210 and classified by a differential mobility analyzer (nano-DMA, Model 3085, TSI). The neutralized silver nanoparticles (by another Po-210) are then introduced to the test filter and the number counts upstream and downstream of the filter are measured by an

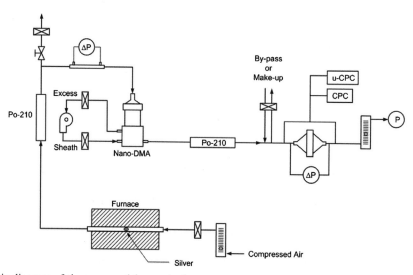

Figure 1. Schematic diagram of the nanoparticle penetration test.

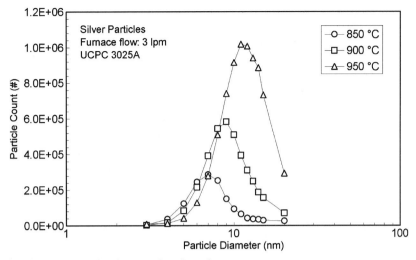

Figure 2. Silver nanoparticle size distribution as a function of temperature.

ultrafine CPC (Model 3025A, TSI) for the nano-particle penetration calculation during a certain sampling time that is long enough to show high reliability. Each test is repeated more than five times with different test conditions to show the variability caused by the set-up and the measurement itself.

Chen et al. (1998) studied nanoparticle transportation in the nano-DMA in order to reduce nanoparticle loss and suggested a new inlet design to reduce the recirculation problem. Here, the slit width is reduced to improve the matching of the flow velocity in the classifying region and to avoid electric field penetration into the upstream side of the entrance slit. As a result, the nano-DMA has the potential for high resolution in sizing and classifying nanoparticles. Figure 3 shows the calibration results of nano-DMA measured by a Scanning Mobility Particle Sizer (SMPS, Model 3080, TSI) and CPC (Model 3022A, TSI). These size distributions were measured next to the nano-DMA for the particle size classification and the results show that nanoparticle size distribution classified by nano-DMA is monodisperse and

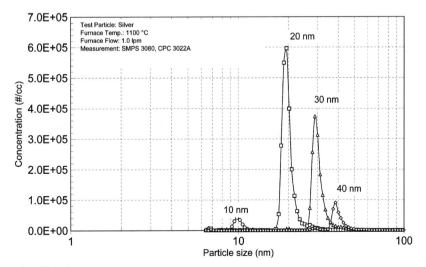

Figure 3. Nano-DMA calibration.

Table 1. Specifications of H&V fiberglass filter media

Filter Parameters		Media			
		HE1073	HE1021	HF0031	HF0012
Thickness (cm)	Ave.	0.053	0.069	0.074	0.074
	%COV	2.3	4.3	2.3	2.3
Basis Weight (g/m^2)	Ave.	63.9	80.3	82.6	69.2
	%COV	0.53	0.67	0.86	0.92
Pressure Drop at 5.3 cm/s (mmH$_2$O)	Ave.	8.4	4.7	3.5	1.3
	%COV	1.48	1.35	1.94	1.47
DOP % Penetration 0.3 μm at 5.3 cm/s	Ave.	12.8	39	45.8	79.9
	%COV	2.2	1.7	0.92	1.24
Fiber Density(g/m^3)	–	2.4	2.4	2.4	2.4
Solidity	–	0.050	0.049	0.047	0.039
Effective Fiber Diameter (μm)	–	1.9	2.9	3.3	4.9
Effective Pore Diameter (μm)	–	8.8	13.4	16.1	26.2

acceptable for a discrete nanoparticle penetration test.

Table 1 shows the specifications of the four different fiberglass filter media tested in this study. The filter media were made by the manufacturer, Hollingsworth and Vose of East Walpole, MA 02032, U.S.A, and were originally donated for use in the establishment of the precision and accuracy statement for the ASTM F1215-89 Standard "Standard Method for Determining the Initial Efficiency of a Flatsheet Filter Media in an Airflow Using Latex Spheres". The filter media are of very low variability, with coefficients of variation for thickness, mass per area, initial pressure drop and initial DOP penetration of less than 4, 1, 2 and 3%, respectively (Japuntich, 1991). The HE series approaches HEPA regime for small particle size and the HF series is more common to standard HVAC systems. These filter media have different combinations of supporting fibers to keep filter shape and main fibers to capture particles, and the filtration efficiency is proportional to the amount of the main fibers. Figure 4 shows the SEM images of H&V fiberglass filter media magnified 500 times. The pore size of the HE filter media is much smaller than that of the HF filter media, and the main fiber ratio of the HE filter media is much higher than that of the HF filter media.

Table 2 shows a list of commercial filter media that were tested in this study. Four different electret filter media (media A, B, C, and D) are made by the 3M Company and Lydall, Inc. and applied on commercial respirators that are widely used in the working field. Media E is a nanosized e-PTFE (expanded polytetrafluoroethylene) membrane filter medium made by W.L. Gore, and is used for ultra high efficiency filtration industrial applications. Figure 5 shows the SEM images of the commercial filter media tested in this study. The uniformity of the porosity is not as good when compared to the H&V fiberglass filter media as shown in the SEM images, but media E can be expected to show high repeatability in test results due to its uniform porosity all over the filter medium.

Each filter sample was placed in a portable filter holder with a filtration area of 17.34 cm^2, and the face velocity through the test filter medium was controlled by a regulated vacuum pump located at

Table 2. Specifications of the specialized filter media

Name	Type	Manufacturing method
Media A	Corona charged blown fiber (mid-size fiber)	Melt blowing process
Media B	Highly charged blown Fiber (mid-size fiber)	Melt blowing process
Media C	Split film fiber	Film extrusion process
Media D	Highly charged blown fiber (fine-size fiber)	Melt blowing process
Media E	e-PTFE Membrane Filter	–

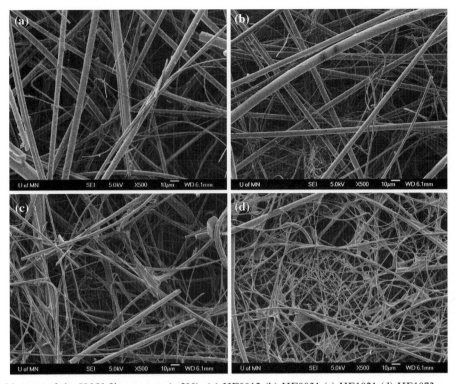

Figure 4. SEM images of the H&V filter papers (×500). (a) HF0012 (b) HF0031 (c) HE1021 (d) HE1073.

the end of the system. The tests were conducted with face velocities of 5.3, 10.0, and 15.0 cm/s. Prior to each measurement, a particle count measurement at the downstream end of the test filter with applying zero Volts to the nano-DMA was conducted in order to check the leakage of the filter holder. Because no particle can pass through the DMA without any electric field inside of the DMA, we can check a system leakage by checking the zero particle count in case of zero Volts DMA test. Furthermore, after switching the sampling point from upstream to downstream, another zero Volts DMA test was conducted to make sure that there are no residual particles inside the sampling tube.

Results and discussion

Nanoparticle penetration efficiencies were measured for four different fiberglass filter media, four different electret filter media and one nanofiber filter medium using silver nanoparticles. All experimental results are shown in terms of the percent penetration with respect to the electrical mobility diameter that is classified by the nano-DMA. Each data point is an average of at least five replicates with the maximum and minimum values as error bars. Figure 6 shows the nanoparticle penetration of the H&V fiberglass filter media at the face velocity of 5.3 cm/s, which is a standard test velocity for a respirator filter medium. The furnace setting temperature was 870°C, which can generate an adequate amount of silver nanoparticles for the size range of 3 to 20 nm. The particle sampling time was 600 s for particle sizes smaller than 5 nm and 60 s for the rest of particle size in order to get more than 10^5 counts at the upstream and 10 counts at the downstream sampling point so that 99.99% efficiency can be detected with the high efficiency filter. The results show very high uniformity with small error bars. In the case of the HF 0012, which has the lowest filtration efficiency among the H&V fiberglass filter media, data were obtainable for all particle sizes down to 3 nm, while the particle

Figure 5. SEM image of the specialized filter media. (a) Media A (×100) (b) Media B (×100) (c) Media C (×50) (d) Media D (×100) (e) Media E (×3,000) (f) Media E (×30,000).

penetration less than 9 nm for the HE 1073 could not be measured due to its high filtration efficiency. In these cases, particles could not be detected at the downstream sampling point, even with an extended sampling time of 30 min. The results show that particle penetration decreases continuously down to 3 nm as expected from the classical filtration theory, and there is no significant evidence of the nanoparticle thermal rebound down to 3 nm.

Figures 7 and 8 show the nanoparticle penetration of the H&V fiberglass filter media at the face velocity of 10 and 15 cm/s, respectively. The higher face velocities show a higher penetration percentage due to a shorter residence time through the filtration region. These results show the same trend as the case of the 5.3 cm/s face velocity. Figure 9 shows the nanoparticle penetration of the commercial filter media at the face velocity of 5.3 cm/s. Nanoparticle penetration decreases

123

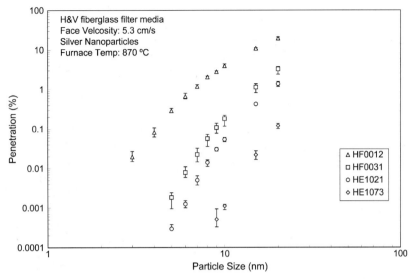

Figure 6. Nanoparticle penetration of the H&V filters at the face velocity of 5.3 cm/s.

continuously with decreasing particle size down to 3 nm with no evidence of thermal rebound. These results show larger error bars than those of the H&V fiberglass filter media due to the non-uniformity of the fiber diameter, porosity and fiber charging condition except media E as mentioned previously.

Figure 10 shows the combination of the nanoparticle penetration measured in this study with the submicron particle (from 20 to 200 nm) penetration measured by Japuntich et al. (2007) for H&V fiberglass filter media at the face velocity of 5.3 cm/s. They used a TSI 8160 automated filter tester for the particle penetration test with sodium chloride (NaCl) particles generated by an atomizer. As shown in the graph, the results agree well with each other at the particle size of 20 nm, even though different test particles were used in the

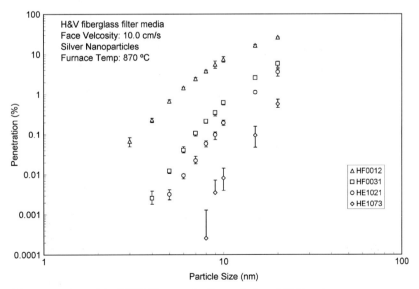

Figure 7. Nanoparticle penetration of the H&V filters at the face velocity of 10.0 cm/s.

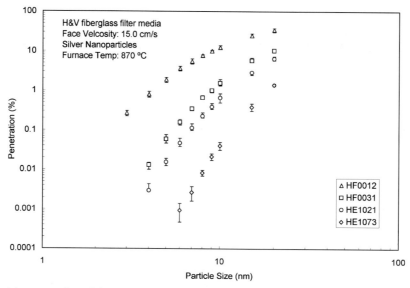

Figure 8. Nanoparticle penetration of the H&V filters at the face velocity of 15.0 cm/s.

two studies. This is because the most dominant filtration mechanism for nanoparticles is Brownian diffusion, which is not affected by the particle density.

Conclusion

In this study, the nanoparticle penetration was tested with a wide range of filter media (four glassfiber filter media, four electret filter media and one nanofiber filter medium) using silver nano-particles from 3 nm to 20 nm at face velocities of 5.3, 10 and 15 cm/s. Nano-DMA calibration and adequate leakage tests show that the test system can produce repeatable and reliable data. The furnace setting temperature and particle sampling time were determined experimentally in order to generate enough amounts of silver nanoparticles in the size range of 3–20 nm, so that 99.99% of effi-

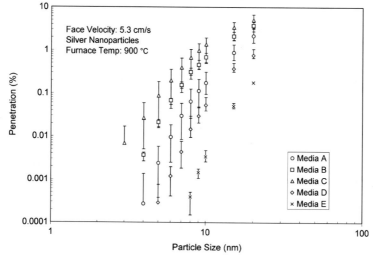

Figure 9. Nanoparticle penetration of the specialized filter media.

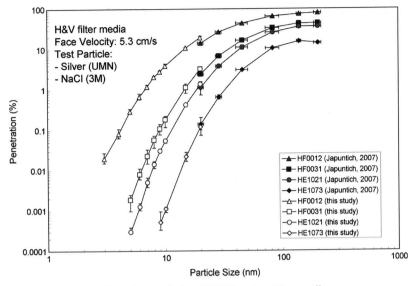

Figure 10. Comparison of test results with other study for H&V fiberglass filter media.

ciency can be measured for the high efficiency filter media. The results show a very small variability with small error bars for all filter media tested in this study, even though each test was repeated many times with a variety of test conditions (operator, test date and sample). The particle penetration decreases continuously down to 3 nm as expected from the classical filtration theory, and there is no significant evidence of nanoparticle thermal rebound down to 3 nm for nine different filter media with three different face velocities. Further, the result shows a good agreement in the overlapping size range of previous test results using submicron particles (from 20 to 200 nm).

Acknowledgement

The authors thank the support of members of the Center for Filtration Research (CFR): 3M, Donaldson, Fleetguard, Samsung Digital Appliance, Samsung Semiconductor, TSI, and W. L. Gore & Associates.

References

Alonso M., Y. Kousaka, T. Hashimoto & N. Hashimoto, 1997. Penetration of nanometer-sized aerosol particle through wire screen and laminar flow tube. Aerosol Sci. Technol. 27, 471–480.

Chen D.-R., D.Y.H. Pui, D. Hummes, H. Fissan, F.R. Quant & G.J. Sem, 1998. Design and evaluation of a nanometer aerosol differential mobility analyzer (Nano-DMA). J. Aerosol Sci. 29(5/6), 497–509.

Hinds W.C., 1999 Aerosol Technology: Properties, Behavior, and Measurement of Airborne Particles. New York: John Wiley & Sons.

Ichitsubo H., T. Hashimoto, M. Alonso & Y. Kousaka, 1996. Panetration of ultrafine particles and ion clusters through wire screen. Aerosol Sci. Technol. 24, 119–127.

Japuntich D.A., 1991 Particle Clogging of Fibrous Filters, Ph.D. Thesis. Loughborough, U.K: Loughborough University of Technology.

Japuntich, D.A., L. Franklin, D.Y.H. Pui, T. Kuehn & S.C. Kim, 2007. A comparison of two nano-sized particle air filtration tests in the diameter range of 10 to 400 nm, J. Nanoparticle Res.

Ku B.K. & A. Maynard, 2006. Generation and investigation of airborne silver nanoparticle with specific size and morphology by homogeneous nucleation, coagulation and sintering. J. Aerosol Sci. 37(4), 452–470.

Maynard A.D., 2003. Estimating aerosol surface area from number and mass concentration measurement. Ann. Occupational Hygiene 47, 123–144.

Oberdörster G., 2000. Toxicology of ultrafine particles: in vivo studies. Phil. Trans. Roy. Soc. London Ser. A 358, 2719–2740.

Wang H.-C. & G. Kasper, 1991. Filtration efficiency of nanometer-size aerosol particles. J. Aerosol Sci. 22, 31–41.

Journal of Nanoparticle Research (2007) 9:127–136
DOI 10.1007/s11051-006-9181-7

© Springer 2006

Special issue: Nanoparticles and Occcupational Health

Reduction of nanoparticle exposure to welding aerosols by modification of the ventilation system in a workplace

Myong-Hwa Lee[1,2], William J. McClellan[1], Joe Candela[3], Dan Andrews[3] and Pratim Biswas[1,*]

[1]*Aerosol and Air Quality Research Laboratory, Department of Energy, Environmental and Chemical Engineering, Washington University in St. Louis, Campus Box 1180, St. Louis, MO, 63130, USA;* [2]*Environment and Energy Division, Korea Institute of Industrial Technology, 35-3, Hongcheon-ri, Ipjang-myeon, Cheonan-si, South Korea;* [3]*Joint Apprenticeship and Training Program, Sheet Metal Workers, St. Louis, MO, 63103, USA;* *Author for correspondence (Tel.: + 1-314-9355482; Fax: + 1-314-9355464; E-mail: pratim.biswas@wustl.edu)*

Received 11 September 2006; accepted in revised form 17 September 2006

Key words: welding aerosols, particle size distribution, nanoparticles, ventilation, occupational health

Abstract

Nanometer particle size distributions were measured in booths with two different ventilation patterns in an occupational environment with welding operations underway. The measurements were used to illustrate the impact of change of ventilation methods (existing – with ventilation ducts located at the top, modified – with ventilation ducts located below the weld bench) on the aerosol size distributions at different locations: close to the weld, in the vicinity of the welder's face, and in the exhaust duct. Particle number concentrations measured in the vicinity of the welder's face (mask) during a horizontal standard arc welding process in a booth with ventilation at the top was in the range of 7.78×10^5 particles cm^{-3} with a geometric mean size of 181 nm and geometric standard deviation of 1.8. This reduced to 1.48×10^4 particles cm^{-3} in the vicinity of the welder's face with the modified ventilation system. The clearance of the welding aerosol was also faster in the modified booth (6 min compared to 11 min in a conventional booth). Particles were collected in the booth for the various test conditions, and analyzed to determine their composition and morphology. The particles were composed of hazardous heavy metals such as manganese, chromium and nickel, and had varying morphologies.

Introduction

Nanoparticles are ubiquitous in the environment as they are emitted from several sources and also formed by spontaneous nucleation in the atmosphere. Nanoparticles in the workplace are of interest, and Biswas and Wu (2005) have reviewed a variety of industrial processes that are sources of unwanted nanoparticles. Vincent and Clement (2000) categorized ultrafine particles in the workplace as: (1) fumes from hot processes (e.g. smelting, refining and welding); (2) fumes from

(incomplete) combustion processes (e.g. transportation, carbon black manufacture); and (3) bioaerosols (e.g. viruses, endotoxins). Conditions required for the generation of ultrafine particles in a workplace were summarized as: (1) presence of vaporizable material; (2) sufficiently high temperature to produce enough vapor, followed by condensation to form an independent aerosol; (3) rapid cooling and a large temperature gradient. Welding is a commonly used industrial process in which all the above conditions are observed, and well known to produce large concentrations of nanoparticles. There are many types of welding processes (Zimmer & Biswas, 2001); and one of the commonly used ones is the arc welding process.

More than 700,000 people are involved with welding for some part of their work, and it has been estimated that up to two percent of the working population in industrialized countries are welders (NIOSH, 1988). Welding processes have been linked to several detrimental acute and chronic health effects (NIOSH, 1988), one reason probably due to exposure to nanoparticles with compositions of chemicals contained in the weld electrode and base metal. Acute health effects include bronchitis, pneumonitis and metal fume fever (McNeilly et al., 2004). Studies have linked lung cancer, cancer of the larynx, and cancer of the urinary tract to welding fume exposure (Stern, 1983; Sjogren et al., 1994; Antonini et al., 2003). Welding fume inhalation can also cause chronic bronchitis, emphysema and increased risk of heart disease. Welding fumes also have non-respiratory effects such as eye and skin irritation (AFSCME, 2004). There have been only limited studies conducted that have characterized aerosols generated from a welding process (Zimmer & Biswas, 2001; Konarski et al., 2003; Stephenson et al., 2003). These studies have shown that a high concentration of nanoparticles are generated in such processes. While results on their capture and removal have been conjectured, the overall concentration of nanoparticles in work places has not been monitored. The proper use of ventilation systems can also help in reducing exposure.

There is increasing interest in establishing the impacts of nanomaterials in workplaces, especially occupational settings wherein engineered nanoparticles are synthesized. While welding processes are not used for deliberate production of nanomaterials, there are several commonalities that are important. First, arc welding is a plasma process that is similar to high temperature processes used to synthesize nanoparticles. Second, it is conducted in an open environment, thus providing valuable data for potential control methods that would be protective of the workers. Third, the unwanted nanoparticles that are produced have to be controlled and captured to minimize exposure to the worker, and may contain valuable by-products which could be a source of useful nanomaterials. This paradigm of reducing emissions and exposure, and producing valuable by-products is a useful approach that should be considered (Biswas et al., 1998).

The objective of this paper is to characterize the impact of ventilation changes on resultant welding aerosol size distributions in an occupational environment. Nanoparticle size distributions are measured in an operator training school wherein arc welding is carried out in a series of booths. Modifications in ventilation systems are made to study the potential impact of nanoparticle exposure. Particle number size distributions, mass concentrations, morphology and chemical compositions are reported. Scenarios that would result in reduced nanoparticle exposure are discussed.

Experimental

The overall objective is to study the influence of ventilation changes on nanoparticle aerosol characteristics in a booth where welding operations are underway. A welding operator training school was used as the model case in this study. A brief description of the booth and welding operation, and the various measurement systems is provided.

Ventilation systems

Welding processes indoors is often done in a small booth consisting of a table, an apparatus to assist with vertical welding processes and other ancillary electrical equipment. For this study two different booths were used, each equipped with different types of ventilation systems as shown in Figure 1. The first was an existing welding booth (conventional booth; 2.1 m high×1.5 m wide×1.8 m deep) with the ventilation duct located overhead at the top in the ceiling. The exhaust duct was a single pipe (diameter 15 cm) and had a flow rate of 11 m^3 min^{-1}. The booth had a curtain on the open side to protect people in the vicinity not involved in the welding from harmful ultraviolet light. The second booth was one with a modified ventilation system (modified booth). The welding table had an integrated vent for use during horizontal welding operations, as shown in Figure 1. The vent drew flow through a set of grates away from the weld region at a flow rate of 11 m^3 min^{-1}. The modified booth was also equipped with a movable ventilation duct, referred to as the arm. The main use for this apparatus was to limit exposure during vertical welding operations. The

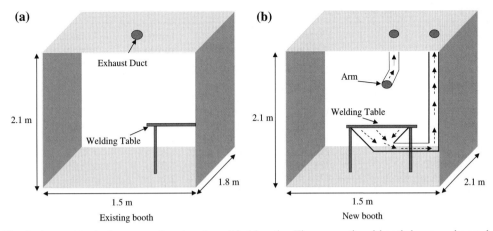

Figure 1. Ventilation system in the conventional and modified booths. The conventional booth has an exhaust duct located overhead. The modified booth ventilation system is below the welding table and has an arm-type exhaust.

diameter of the duct that made up the arm was 14 cm, and operated at a suction flow rate of 9 m^3 min^{-1}.

An arc welding operation on 3/8″ base metal plates (ASTM A36-97a) was carried out (Arc voltage: 20 V, Arc current: 125 A) with E-7018 welding rods (two rods per sample). Both horizontal and vertical welding operations were conducted. The total welding operation ran for about 5 min, allowing sufficient time for sampling using the real time aerosol instruments.

Particle size distribution sampling

To obtain particle size distribution data, a real-time size-distribution measuring system (Scanning Mobility Particle Sizer (SMPS), TSI Inc., St.Paul, MN, USA) was used. This device consists of an electrostatic classifier (TSI model 3080) and a condensation particle counter (TSI model 3025). The sample flow rate was 0.3 liters per minute and the sheath flow rate for the SMPS was 10 liters per minute. The length of the sampling tube was 2.7 meters, so that the operator could conduct the welding operations without disturbance. The loss of particles by diffusional deposition in the tube was not very high, for example about 6% of 50 nm particles would be lost. Prior to the beginning of the welding operation, background measurements were taken to ensure that the concentrations were typical of an indoor environment without any sources. Measurements were taken at different locations (see Table 1) for both horizontal and vertical welding operations. In the conventional

booth, measurements were taken at three different locations: 10 cm above the weld, in the welder's mask near the mouth and nostrils, and in the exhaust duct. The same was done for the booth with the modified ventilation system (modified booth). Temporal measurements were also taken in the booth to determine the time it took for the particle concentrations to drop back to the background levels after welding operations. Replicate tests were conducted and at least three measurements were made for each sampling point.

Particle collection for mass concentration, composition and morphology

Particles from the welding fumes were collected on a 47 mm Teflon filter that was installed in a holder that was 1.5 m above the welding region. The flow rate used was 10 liters per minute, and sampling was done for about 5 min that matched the welding operation. Three replicate filter samples were collected, and the mass of particles collected was determined gravimetrically to determine the mass concentration. The particles were analyzed by X-Ray Fluorescence (XRF, Chester Labnet, Tigard, OR, USA) to determine the elemental concentrations of the constituents in the welding aerosol. The collected particles were also viewed by scanning electron microscopy (SEM, Hitachi S-4500). A trap consisting of a permanent magnet outside the sampling tube was used to determine the fraction of particles that were magnetic, and to check their morphology.

Table 1. Summary of particle size distribution measurements

Sampling location		Particle size, D_p (nm)	Total particles, N_{tot} (#/cm^3)	Geometric standard deviation, σ_g	Total surface area, S_{tot} (nm^2/cm^3)	Total volume, V_{tot} (nm^3/cm^3)
Conventional booth						
Background		52	9.08E+03	2.1	2.48E+08	7.11E+09
Horizontal	Above Weld	162	2.06E+06	1.7	2.62E+11	9.81E+12
weld	In Mask	181	7.78E+05	1.8	1.14E+11	4.44E+12
	In Exhaust Duct	187	2.77E+05	1.6	3.94E+10	1.53E+12
Vertical weld	Above Weld	216	1.69E+05	1.4	2.82E+11	1.13E+13
	In Mask	138	3.29E+05	2.2	3.71E+10	1.43E+12
	In Exhaust Duct	132	2.29E+05	2.1	2.88E+10	1.10E+12
Modified booth						
Background		41	1.26E+04	2.2	2.42E+08	6.34E+09
Horizontal	In Mask (fan off)	188	4.26E+05	1.6	6.18E+10	2.41E+12
weld	In Mask (fan on)	50	1.48E+04	2.1	4.09E+08	1.26E+10
	Below Table	190	4.86E+06	1.6	6.82E+11	2.61E+13
	In Exhaust Duct	159	1.80E+06	1.6	1.96E+11	4.56E+23
	Above Weld	98	2.38E+04	2.4	2.21E+09	8.19E+10
Vertical	In Mask (table only)	116	5.25E+04	2.4	5.95E+09	2.29E+11
Weld	In Mask (table and arm)	45	2.58E+04	2.3	6.88E+08	2.25E+10
	In Mask (arm only)	46	2.12E+04	2.3	5.27E+08	1.58E+10
	Arm Duct	72	2.69E+05	2.1	1.22E+10	3.55E+11

Results and discussion

Particle formation and transport

Welding aerosol formation and the resultant particle transport are complex phenomena. Metallic species from both the electrodes and the base metal being welded are vaporized in the high temperature zone near the arc. The rate of fume generation is a function of welding parameters such as welding alloy, arc voltage, and electrode (Zimmer & Biswas, 2001; Zimmer et al., 2002), resulting in a highly variable aerosol, both spatially and temporally. Controlled studies of a Gas Metal Arc Welding (GMAC) show that at a height of 4.8 cm, the concentrations were 1.42×10^7 particles cm^{-3} and at a height of 19.2 cm the particle concentration decreased to 3.67×10^6 particles cm^{-3} (Zimmer & Biswas, 2001). Near the welding arc, where the temperatures are still very high, the welding fume remains as a vapor. As it begins to move away from the hot zones of the plasma to the cooler regions, the metal species vapors begin to nucleate, followed by growth by condensation and coagulation.

In contrast to the controlled studies conducted earlier, the objective was to evaluate the particle characteristics at the different locations in the booths with the different ventilation systems. In the conventional booth, the generated particles and vapors follow the flow streamlines towards the exhaust duct located in the ceiling, and undergo growth mechanisms (as illustrated in Figure 2). This flow is past the welder's face and protective hood (mask), thus increasing chances of exposure. In contrast, in the modified booth the particles and vapors are drawn away from the vicinity of the welder's face into the duct below the table, and most of the particle growth occurs inside the exhaust duct. Detailed description of the nanoparticle size distributions measured are discussed in the following section.

Conventional booth

Figures 3(a) and (b) show the particle size distributions at different sampling locations in the horizontal and vertical welding process, respectively. The size distributions are very similar above the weld, in the mask and in the exhaust duct; however, there are clear differences in the total

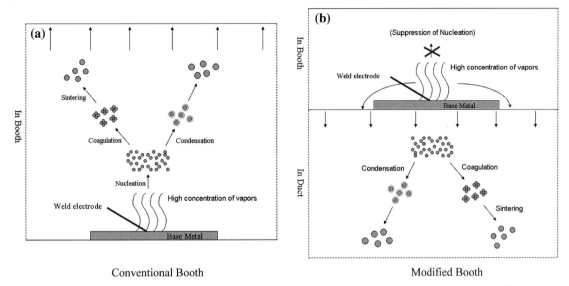

Figure 2. Anticipated particle formation and growth process of welding aerosols in the two booths.

particle number concentrations. The particles typically grow to about 100 to 300 nm very rapidly, as observed in Figure 3. The particles in this range are the most problematic as they deposit deep in the respiratory system and have the longest lung clearance times after deposition (Kuo et al., 2005). Table 1 shows the integral properties of the distribution at the different sampling locations. The number concentrations during welding are considerably higher than that of the background for all particle sizes.

In a horizontal welding operation, the geometric mean size was 181 nm at a number concentration of 7.78×10^5 particles cm^{-3} in the welder's mask. The corresponding mean size in the mask was 52 nm with a number concentration of 9.08×10^3 particles cm^{-3} for the background condition. Closer to the weld, the geometric mean size was smaller (162 nm), and the size correspondingly increased to 181 nm in the mask, and 187 nm in the exhaust duct. This is due to the particle growth by condensation and coagulation mechanisms described earlier. The total number concentrations decrease with distance away from the weld zone: from 2.06×10^6 particles cm^{-3} above the weld, 7.78×10^5 particles cm^{-3} in the mask to 2.77×10^5 particles cm^{-3} in the exhaust duct. The particles in the weld zone were lower than that measured in the controlled studies of Zimmer and Biswas (2001), due to the enhanced mixing and dilution in an actual welding booth environment.

The decrease with distance away from the weld zone is due to both growth by coagulation, and due to dilution as the aerosol plume moves away and spreads. The decrease in the total volume concentration reported in Table 1 confirms that dilution is also playing a role.

The vertical welding operation reveals a different pattern. Samples taken in the vicinity of the welding zone had a higher mean particle size (216 nm) which then decreased for samples taken in the mask (138 nm) and the exhaust duct (132 nm). The total number concentration above the welding region was 1.69×10^5 particles cm^{-3}, and in the mask and at the exhaust duct they were 3.29×10^5 particles cm^{-3} and 2.29×10^5 particles cm^{-3}, respectively. The larger particle sizes above the weld are probably due to the different orientation of the welding process, and complicated flow patterns that affect the growth mechanisms. An important observation is that there was some spatter, and this could also bias the interpretation of the integral properties. Conditions for spatter have been identified in other welding studies (Zimmer et al., 2002).

Modified booth

The modified booth was quite effective at removing the particles in the nanometer size ranges, resulting in a significant decrease in the measured

132

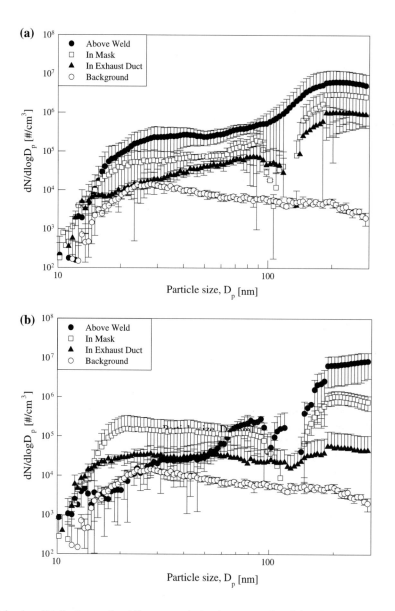

Figure 3. (a). Particle size distributions of welding aerosols in the conventional booth at different locations (Horizontal welding). (b). Particle size distributions of welding aerosols in the conventional booth at different locations (Vertical welding).

total number concentration. Figures 4(a) and (b) show the particle size distributions in the modified booth with the exhaust fan on. Samples taken in the mask without the fan on, yielded a total number concentration of 4.26×10^5 particles cm^{-3} with a particle size of 188 nm as shown in Table 1. With the fan on the total number concentration dropped to 1.48×10^4 particles cm^{-3}. The mean

particle size shifted to a much smaller size (50 nm) with the fan on, closer to background levels in this specific booth.

During a vertical welding operation, the total number concentration in the mask was 5.25×10^4 particles cm^{-3}. This is slightly higher than during a horizontal welding operation, however a major difference is the mean particle size was observed

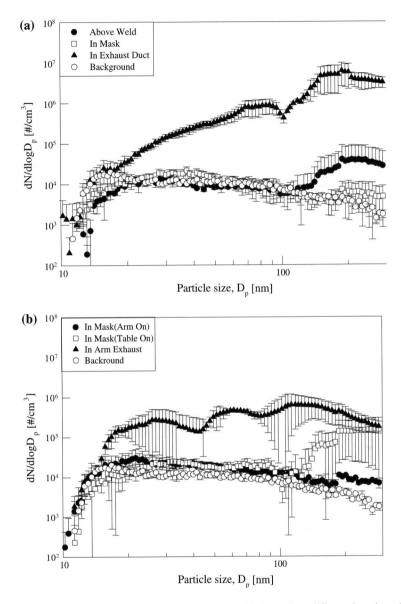

Figure 4. (a). Particle size distributions of welding aerosols in the modified booth at different locations (Horizontal welding). (b). Particle size distributions of welding aerosols in the modified booth at different locations (Vertical welding).

(116 nm for the vertical compared to 50 nm for the horizontal). In a vertical operation, the hood or mask of the welder is closer to the welding zone or region of fume generation that explains the above observed trends.

Using an exhaust duct on the arm allowed more effective drawing away of the generated particles from the vicinity of the welding zone. This resulted

in a drop in the number concentration, and the mean sizes were similar for both types of welding operation (Table 1). The results indicate that the modified ventilation scheme with the exhaust located below the table for the horizontal welding operations, and the arm for the vertical welding operations, are effective in reducing exposure to welders.

134

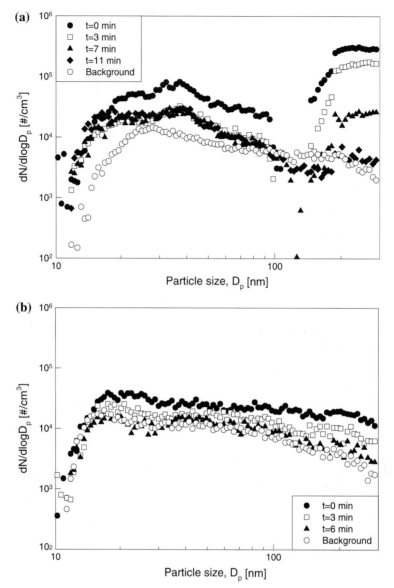

Figure 5. (a) Particle size distributions in the conventional booth for 4 different times after conclusion of welding. (b) Particle size distributions in the modified booth for 3 different times after conclusion of welding.

Particle clearance time in booth

After conclusion of welding, it is important how fast the welding fumes are removed from the booth to protect welders. The time it takes for the aerosol concentrations to reduce down to background levels is shown in Figure 5(a) and (b) for the conventional and modified booths, respectively. The measurements were initiated immediately after the welding process concluded, and taken at the indicated time intervals. The particle concentrations reached background levels in approximately 11 min in the conventional booth, and in 6 min in the modified booth. Thus the efficacy of the modified ventilation system was not only in reducing exposure during the welding operation, but also cleared the environment of the remnant aerosol in a shorter duration.

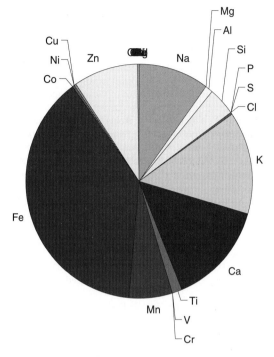

Figure 6. Chemical composition of welding aerosols collected and analyzed by XRF analysis.

Mass concentration, chemical composition and morphology of welding aerosols

The gravimetrically measured mass concentration of collected particles varied between 6.48 and 24.4 mg m^{-3}, and is in the typical range encountered in an occupational environment with welding operations. On assuming an average density for the particles (1.2 g cm^{-3}), and converting the measured number size distributions to the total mass concentration, the values obtained are in the range of 1.4–25.0 mg m^{-3} above the weld region. The gravimetrically measured mass concentration in the booth (6.48–24.4 mg m^{-3}) exceeds the PEL (Permissible Exposure Limit – 5 mg m^{-3}) proposed by ACGIH (1994). The chemical composition of the samples is shown in Figure 6. Several hazardous heavy metal species such as manganese, chromium and nickel are detected in the sample. The composition of the welding aerosol matches that of the welding rod (Fe: 70–85%, CaCO$_3$: 5–10%, CaF$_2$: 5–10, TiO$_2$: 1–3%, Mn: 1.1%, Cr: 0.04%) and the base metal (Fe: 98%, Mn: 0.8%, Cu: 0.5%, Cr: 0.2%, Ni: 0.1% Others: 0.4%). The morphologies of the collected particles are shown

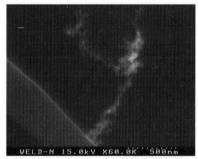
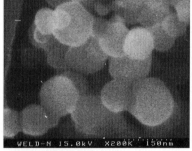

(a) Particles collected on the filter

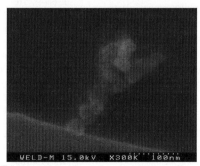

(b) Particles collected on a magnetic filter

Figure 7. Scanning electron microscope micrographs of representative welding aerosols.

in Figure 7. Particles of varying morphology were obtained. Some of them formed more compact agglomerated structures, indicative of a partial sintering process. Others formed more chain like aggregates that are probably indicative of structures formed when magnetic particles grow by collisional processes. To confirm that the particles were magnetic, a filter consisting of an externally imposed magnetic field was used. This selectively collected the ferromagnetic particles, and a linear chain like structure was observed, that matched the morphology of some of the particles collected on the filter. Nanostructured magnetic particles have application in a variety of fields, including that of environmental remediation (Savage & Diallo, 2005). Hence, in addition to capture of nanoparticles to protect the worker, systems using magnetic fields could be used to separate byproducts for use in other applications. More studies need to be conducted before such techniques would be feasible; however they offer potential of reducing emissions and producing valuable nanomaterial by-products.

Conclusions

Welding processes generate nanoparticles and high concentrations are observed in the vicinity of the face of the welder. The particle number concentration from a horizontal standard arc welding process was approximately 7.78×10^5 particles cm^{-3}, with a geometric mean size of 181 nm and geometric standard deviation of 1.8. The mass concentrations varied from 6.48 to 24.4 mg m^{-3}, much higher than the permissible exposure limits. These particles had hazardous constituents and were observed with varying morphologies. On changing the ventilation system in the booth, the concentrations were lower (1.48×10^4 particles cm^{-3}), and also resulted in faster clearance of the generated particles (6 min compared to 11 min).

The studies provide guidance to design of effective ventilation systems to aid in the reduction of nanoparticle concentrations and reduction of exposure. Detailed studies that map out flow patterns would be essential to design systems to minimize airborne concentrations.

References

AFSCME, 2004. Welding hazards. http://www.afscme.org/health/faq-weld.htm.

ACGIH, 1994. Threshold limit values for chemical substances and physical agents and biological exposure indices, Cincinnati, OH. American Conference of Governmental Industrial Hygienists.

Antonini J.M., A.B. Lewis, J.R. Roberts & D.A. Whaley, 2003. Pulmonary effects of welding fumes: Review of worker and experimental animal studies. Am. J. Ind. Med. 43(4), 350–360.

Biswas P., G. Yang & M.R. Zachariah, 1998. In situ processing of ferroelectric materials from lead waste streams by injection of gas phase titanium precursors: Laser induced fluorescence and X-ray diffraction measurements. Combust. Sci. Technol. 134(1–6), 183–200.

Biswas P. & C.Y. Wu, 2005. Nanoparticles and the environment: A critical review paper. J. Air Waste Manage. Assoc. 55, 708–746.

Konarski P., I. Iwanejko & M. Cwil, 2003. Core-shell morphology of welding fume micro- and nanoparticles. Vacuum 70(2–3), 385–389.

Kuo Y.M., S.H. Huang, T.S. Shih, C.C. Chen, Y.M. Weng & W.Y. Lin, 2005. Development of a size-selective inlet-simulating ICRP lung deposition fraction. Aerosol Sci. Technol. 39(5), 437–443.

McNeilly J.D., M.R. Heal, I.J. Beverland, A. Howe, M.D. Gibson, l.R. Hibbs, W. MacNee & K. Donaldson, 2004. Soluble transition metals cause the pro-inflammatory effects of welding fumes in vitro. Toxicol. Appl. Pharmacol. 196(1), 95–107.

NIOSH, 1988. *NIOSH* Criteria for a Recommended Standard: Welding, brazing, and Thermal Cutting. DHHS (NIOSH) publication No. 88–110, Cincinnati, OH.

Savage N. & M.S. Diallo, 2005. Nanomaterials and water purification: Opportunities and challenges. J. Nanoparticle Res. 7(4–5), 331–342.

Sjogren B., K.S. Hansen, H. Kjuus & P.G. Persson, 1994. Exposure to stainless-steel welding fumes and lung-cancer – a metaanalysis. Occup. Environ. Med. 51(5), 335–336.

Stephenson D., G. Seshadri & J.M. Veranth, 2003. Workplace exposure to submicron particle mass and number concentrations from manual arc welding of carbon steel. AIHA J. 64(4), 516–521.

Stern R.M., 1983. Assessment of risk of lung cancer for welders. Arch. Environ. Health 38, 148–155.

Vincent J.H. & C.F. Clement, 2000. Ultrafine particles in workplace atmospheres. Phil. Trans. Royal Soc. London A Mathematical Phys. Eng. Sci. 358(1775), 2673–2682.

Zimmer A.T., P.A. Baron & P. Biswas, 2002. The influence of operational parameters on number-weighted aerosol size distribution generated from a gas metal arc welding process. J. Aerosol Sci. 33, 519–531.

Zimmer A.T. & P. Biswas, 2001. Characterization of the aerosols resulting from arc welding processes. J. Aerosol Sci. 32, 993–1008.

Journal of Nanoparticle Research (2007) 9:137–156
DOI 10.1007/s11051-006-9154-x

Special focus: Nanoparticles and Occupational Health

Health risk assessment for nanoparticles: A case for using expert judgment

Milind Kandlikar[1], Gurumurthy Ramachandran[2,*], Andrew Maynard[3], Barbara Murdock[2]
and William A. Toscano[2]
[1]Institute for Resources, Environment and Sustainability, University of British Columbia, Vancouver, Canada;
[2]Division of Environmental Health Sciences, School of Public Health, University of Minnesota, Minneapolis,
MN, USA; [3]Woodrow Wilson International Center for Scholars, The Smithsonian Institution,
Washington, DC, USA; *Author for correspondence (Tel.: +1-612-626-5428; Fax: +1-612-626-4837;
E-mail: ramac002@umn.edu)

Received 24 July 2006; accepted in revised form 2 August 2006

Key words: nanoparticle health risks, deep uncertainty, parametric uncertainty, model uncertainty,
probabilistic expert judgment, degree of expert consensus, occupational health

Abstract

Uncertainties in conventional quantitative risk assessment typically relate to values of parameters in
risk models. For many environmental contaminants, there is a lack of sufficient information about
multiple components of the risk assessment framework. In such cases, the use of default assumptions
and extrapolations to fill in the data gaps is a common practice. Nanoparticle risks, however, pose a
new form of risk assessment challenge. Besides a lack of data, there is deep scientific uncertainty
regarding every aspect of the risk assessment framework: (a) particle characteristics that may affect
toxicity; (b) their fate and transport through the environment; (c) the routes of exposure and the
metrics by which exposure ought to be measured; (d) the mechanisms of translocation to different parts
of the body; and (e) the mechanisms of toxicity and disease. In each of these areas, there are multiple
and competing models and hypotheses. These are not merely parametric uncertainties but uncertainties
about the choice of the causal mechanisms themselves and the proper model variables to be used, i.e.,
structural uncertainties. While these uncertainties exist for PM2.5 as well, risk assessment for PM2.5
has avoided dealing with these issues because of a plethora of epidemiological studies. However, such
studies don't exist for the case of nanoparticles. Even if such studies are done in the future, they will be
very specific to a particular type of engineered nanoparticle and not generalizable to other nanoparticles. Therefore, risk assessment for nanoparticles will have to deal with the various uncertainties that
were avoided in the case of PM2.5. Consequently, uncertainties in estimating risks due to nanoparticle
exposures may be characterized as 'extreme'. This paper proposes a methodology by which risk analysts
can cope with such extreme uncertainty. One way to make these problems analytically tractable is to
use expert judgment approaches to study the degree of consensus and/or disagreement between experts
on different parts of the exposure–response paradigm. This can be done by eliciting judgments from a
wide range of experts on different parts of the risk causal chain. We also use examples to illustrate how
studying expert consensus/disagreement helps in research prioritization and budget allocation exercises.
The expert elicitation can be repeated over the course of several years, over which time, the state of
scientific knowledge will also improve and uncertainties may possibly reduce. Results from expert the
elicitation exercise can be used by risk managers or managers of funding agencies as a tool for research
prioritization.

Introduction

The appeal of nanotech results from its protean nature – applications of this technology have the potential to affect all aspects of human life. Worldwide government investment in nanotechnology has increased by a factor of five from $825 million in 2000 to $4.1 billion in 2005 (Roco, 2005). However, like biotechnology before it, concerns about environmental and human health risks have already begun to have an effect on the societal debate around nanotechnology. Fears about risks of nanotechnology result from a basic conundrum: the properties that make nanoparticles so promising – that they can behave very differently from bulk forms of the same material – also make their health and environmental effects extremely difficult to predict (Service, 2004). Uncertainties about health effects feed directly into the risk-benefit debates that increasingly shape societal responses to, and regulation of, new technologies.

The human health impact of toxic substances and pollutants can be studied using frameworks of risk assessment developed over the past 30 years. Risk assessment in this context is a set of tools used to integrate exposure and health effect information for characterizing the potential for health hazards to humans (NRC/NAS, 1983; US EPA, 2004). Such methods typically use quantitative predictions of health impacts (e.g., probability of mortality). However, qualitative risk assessments are also valuable when quantitative assessments are not possible. Even under the best of circumstances, risk assessment cannot estimate risk with absolute certainty. Modern quantitative risk assessment aims not to arrive at a single precise number, but to allow decision makers to face the possible consequences of a range of "not clearly incorrect" answers and decide on the protective policies that are warranted in light of the range of possible future outcomes of alternative policies (Anderson & Hattis, 1999). Thus, for example, if there is uncertainty regarding exposures or dose in a population, one can either collect more data or use numerical models to estimate missing values or extrapolate values from other similar populations.

Numerical models incorporate amounts of pollutants prevalent in the ambient environment, the actual amounts absorbed or inhaled by an exposed individual, the toxicological effects on the human body as inferred from *in vivo*, *in vitro*, *in silico*, and epidemiological studies, and biologic variability. Uncertainties in values of model parameters play a central role in conventional risk calculations. The parameters in such models may be known with some uncertainty, and a degree of subjective professional judgment about the values of such parameters may enter into the analysis. Uncertainties in numerical models such as those used to calculate risks can be divided into two broad classes: (a) "parametric" uncertainty associated with parameter or observational values that are not known precisely and (b) "structural" or "model" uncertainty where important relationships between variables or their functional form may not have been identified correctly. Not surprisingly, assessing model uncertainty is generally more difficult since the uncertainties are likely to be more fundamental. Since key mechanisms for exposure processes and toxicity effects of manufactured nanoparticles remain poorly understood, model uncertainties might dominate in related risk calculations. Uncertainties about mechanism include those related to: (a) how long manufactured nanoparticles may persist in the atmosphere depending on their rates of agglomeration (and some nanoparticles are designed specifically not to agglomerate), thus influencing the probability of exposure; (b) the routes of exposure and the metrics by which exposure ought to be measured; (c) mechanisms of translocation to different parts of the body after nanoparticles enter the body; (d) mechanisms of toxicity. These are not simply uncertainties in the value of some model parameter but rather uncertainties about the causal mechanisms themselves.

This paper examines how risk analysts can cope with extreme uncertainty in nanoparticle risk calculations. In Section "Traditional risk assessment" we begin with a discussion of uncertainties in health risks from particles that are less than 2.5 μm in diameter ($PM_{2.5}$). Risks from $PM_{2.5}$ have been studied for over a decade and provide a starting point for understanding the health effects of particulate matter. In Section "Uncertainty in characterizing health risks from nanoparticles", we extend the analysis to manufactured nanoparticles and show how model uncertainties dominate

139

calculation of risks. In Section "Expert judgment in uncertainty assessment", we suggest that one way to address the presence of extreme uncertainty is to use expert assessment as a tool for assessing uncertainties. We show how systematic use of expert judgment approaches might help in developing an understanding of the aspects of the problem that are well understood and those that are not. We also use examples to illustrate how studying expert consensus/disagreement helps in research prioritization. We conclude by presenting some thoughts on the challenges of using expert assessment as tool for analyzing risks from manufactured nanoparticles (Section "Conclusions").

Traditional risk assessment

Risk assessment is a complex process that involves the integration of information across a range of domains including source characterization, fate and transport, modeling, exposure assessment, and dose–response characteristics. It uses well-defined quantitative models to describe the relationships

between the various elements of the paradigm shown in Figure 1. We begin by briefly reviewing how health risks have traditionally been identified and quantified based on information about exposure and dose–response relationships. Implicit in this process is the setting of "standards" or guidelines regarding "safe" or "acceptable" levels of exposure for a population. Figure 1 shows the general environmental health framework (in the center) and its relationship to the risk assessment framework (loosely based on Sexton et al., 1995). Exposure is defined as the intensity of contact between contaminant and the relevant biological sites of impact over a relevant time period. Exposure assessment includes assessing sources of pollutants and their strengths, measuring or modeling concentrations in environmental media, measuring or modeling human exposures through various pathways, and in some cases even biological monitoring to measure tissue burden and thereby estimate dose. The estimation of a biologically relevant dose from exposure information is, however, often very difficult and requires fairly detailed knowledge of the toxicokinetics of the pollutant in the human body.

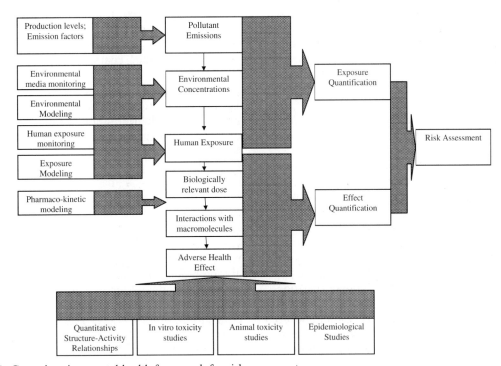

Figure 1. General environmental health framework for risk assessment.

The lower portion of the diagram relates to the estimation of the health effects of the exposure and biologically relevant dose. Information regarding effects can come from *in vitro* and *in vivo* studies, quantitative structure–activity relationship (QSAR) modeling (Coleman et al., 2003), and epidemiological studies. The quantification of exposures and effects allows the quantification of risk that, in turn, allows the proper allocation of resources to manage the risk. For example, risk assessment might allow the identification of populations or individuals at greater risk because their exposures are greater than some threshold identified in epidemiological studies.

For non-cancer toxicants, it is often assumed that there is some level below which there are no adverse effects [no observed adverse effects level (NOAEL)]. A limit or guideline or reference concentration is established below this threshold. Exposures exceeding this threshold are considered to cause adverse health effects, and measures to mitigate or reduce exposures need to be taken. If the shape of the exposure–response curve above the threshold is well defined, then health risk calculations are feasible. For carcinogens, the standard practice has been to extrapolate from high dose to low dose by assuming a linear dose–response curve without any threshold level (i.e., the threshold is zero). The excess risk is calculated by multiplying the dose with the dose–response curve "slope factor". Even though the model assumes no threshold level, for the purpose of risk management and prioritization, risks exceeding some minimum risk probability (e.g., $1/10^3$ for occupational populations and $1/10^6$ for non-occupational populations) are considered to be of concern.

Uncertainties in the quantification of model can arise due to measurement errors (e.g., errors in measuring emissions from sources), systematic biases (e.g., using centrally located outdoor monitors for air pollutants to estimate exposures for a population that spends most of its time indoors), and non-representativeness (e.g., estimating the slope of an exposure–response curve from an epidemiological study done with a sample that does not represent the general population). In the above examples, the underlying scientific model and the causal chain is not in dispute. Only the values of the parameters in the model are uncertain. In such cases probabilistic methods such as Monte Carlo

analysis can effectively characterize uncertainty. In contrast, model uncertainty arises because the relationships within and among various components of the risk assessment continuum are poorly understood. There may be several competing models that purport to explain the relationships between variables of interest. Competing model forms may have different parameters, may have different functional forms, and may even have mutually exclusive assumptions. In the worst case, there may be no explanatory models at all. Thus, the uncertainty is not so much the value of a parameter in a model, but in the choice of the model itself. Such uncertainties are difficult, if not impossible, to quantify.

The example of $PM_{2.5}$

Some features of the above can be observed in the development of the national standards in the US for $PM_{2.5}$. Since 1987, PM10 has been the metric for measuring airborne particulates for regulatory compliance. The $PM_{2.5}$ fraction that the US EPA also regulates now is a sub-fraction of PM10.

Uncertainty regarding health effects studies
The first strong findings supporting $PM_{2.5}$ health effects were derived from six US cities where daily mortality was associated with short-term increases in $PM_{2.5}$ and only weakly associated with increases in the coarse fraction ($PM_{2.5–10}$) (Schwartz et al., 1996). These findings were supported by subsequent studies (reviewed in US EPA, 1996; Fairley, 1999). However, these and other epidemiological studies could be criticized on several grounds including, (1) use of inappropriate statistical methods, (2) inability to account for meteorology and co-pollutants, exposure misclassification, measurement error, and (3) a lack of biological plausibility (Vedal, 1997). In addition, different investigators provided divergent analyses and interpretations of the same data using different exposure and health metrics. This divergence was most clearly seen in the analyses of the data on air pollution and mortality from Philadelphia made by different investigators (Schwartz & Dockery, 1992; Schwartz, 1994; Moolgavkar et al., 1995; Moolgavkar & Luebeck, 1996; Samet et al., 1995, 1997).

Health effects of PM became persuasive only when evidence from much larger epidemiological studies became available. Samet et al. (2000) investigated the association of PM10 and mortality in the 90 largest US cities and found that a 10 μg/m^3 increase in PM10 resulted in a 0.5% increase in mortality from all causes. Adding other pollutants to the model did not affect this result. Regional differences were observed and were likely related to variations in PM10 composition. This study addressed two criticisms raised about previous time-series studies, namely that statistical methods had varied among studies and that the selection of cities for studies may have been biased which in turn might have led to biased results. The emerging consistency and coherence of results from the observational studies was complemented by experimental animal and human data to elucidate attributes of PM that are causally linked to health effects and associated toxicological mechanisms.

Uncertainty regarding PM characteristics

PM is a complex mixture of solid and liquid particles that shows tremendous variability in size, composition and concentration depending on the source of the particles, geography, and seasonal or diurnal meteorology. The coarser particles within the PM10 category consist of crustal minerals, sea salts and biological material while the finer fraction consists of carbonaceous particles with a variety of metals (e.g., iron, vanadium, nickel, copper, and platinum), ionic species such as sulfates and nitrates, acids, reactive gases adsorbed on the particles, hydrocarbons and other organic material. Each one of these components has been hypothesized to be the cause of toxic or adverse health effects.

Particle size is an important parameter in causing toxicity (e.g., Schwartz et al., 1996; Fairley, 1999; Burnett et al., 2000). As early as 1996, regulatory bodies such as the US EPA had recognized that smaller particles were likely to be more toxic. Wichmann et al. (2000) found that over a 3-year period, both ultrafine (PM < 0.1) and fine particles (PM 0.1–2.5) were associated with increased daily mortality. Lippmann et al. (2000) found that size is more significantly associated with adverse health outcomes than chemical composition of the particle.

In addition to size, the presence of transition metals has been linked to toxicity of the air pollutant. Ghio and Devlin (2001) combined epidemiological and experimental approaches to show that air pollution particles collected at an ambient site in Utah showed greater inflammatory responses in human volunteers when the particles had higher metal content (iron, zinc, copper, lead, and nickel). Other studies have shown that rodents exposed to residual fly ash high in metal content displayed inflammatory, respiratory and cardiovascular responses (Watkinson et al., 1998; Campen et al., 2001). More recently, Hahn et al. (2005) used rats exposed to ultrafine and fine airborne particles of carbon and vanadium pentoxide (V_2O_5) to demonstrate that both particle size and chemical composition may be important for particle toxicity. However, these experiments did not account for differences in particle characteristics that may affect toxicity such as insolubility, porosity, and tendency to aggregate. Thus, while there has been substantial progress in identifying specific particle characteristics responsible for toxicity, a complex combination of characteristics may be responsible making risk assessment a more data intensive task

Uncertainty regarding mechanisms of effect

While human experimental studies in controlled exposure chambers were inconclusive about adverse health effects of fine particles, animal toxicological studies provide support for the biological plausibility of the epidemiological findings (e.g., Dreher et al., 1996). These studies also support the idea that individuals with pre-existing respiratory diseases are more susceptible to actions of fine particle exposure (Campen et al., 1996; Costa et al., 1996; Godleski et al., 1996; Ferin et al., 1992). More recently toxicological studies have focused on early cellular/biochemical events that occur after exposure to particles. The inhaled particles elicit an inflammatory response in the airways measured by an increase in the number of neutrophils and in cytokine and chemokine levels resulting from oxidative stress pathways inside cells (Salvi et al., 1999, 2000; Ghio et al., 2000). The inflammatory response may damage the epithelial cell layer as well as macrophages, leading to decreased phagocytosis, i.e., clearing of the particles from the lungs (Renwick et al., 2001). Neurogenic inflammation can also occur where airway nerve cells synthesize neurotransmitters that may affect white blood cells and epithelial

142

tissue (Barnes, 2001). Such inflammation may reduce respiratory function or cause increased susceptibility to infections. In addition to localized inflammatory responses, there may be systemic responses characterized by circulating neutrophils, fibrinogen, plasma viscosity, platelet numbers, and C-reactive protein (Peters et al., 1997a; Salvi et al., 1999; Seaton et al., 1999; Pekkanen et al., 2000; Ibald-Mulli et al., 2001; Peters et al., 2001; Schwartz, 2001; Vincent et al., 2001). Some of these responses are consistent with causal hypotheses for the health effects of PM, e.g., higher levels of fibrinogen could increase plasma viscosity and blood coagulation leading to an increased formation of blood clots in susceptible individuals. Atherosclerosis, arrhythmias, myocardial infarctions and other health end points can be similarly related to systemic responses. However, many of the studies reach conclusions that are inconsistent with each other and with the causal hypotheses. At the same time, the relationship between localized inflammation and systemic responses is not completely understood. Thus, while there is substantial new information about early biological events after exposure, the relationship of these events to short- and long-term health effects and the mechanisms by which one affects the other are poorly understood.

Uncertainty regarding dose–response relationship
The uncertainty in the dose–response relationship that is the basis for risk assessment results from several factors. The data from epidemiological studies are necessarily "noisy" and establishing a shape for a dose–response curve is difficult. Key uncertainties are whether the data allow for a threshold exposure level, the value of such a threshold, and the choice of a linear or non-linear model. Other uncertainties relate to whether the relevant exposure metric should be the total mass of particles less than 2.5 μm or the mass of a specific chemical constituent, whether the standard should be based on an averaging time of 24 h or 1 year (related to whether acute or chronic effects should drive standards-setting), and whether some groups in the population are more susceptible to the effects of PM pollution (e.g., people with cardiovascular disease or diabetes and age). While the standard proposed and eventually adopted by the US EPA (an annual average of 15 μg/m^3 and a 24-h average of 65 μg/m^3) assumes a threshold, there

is substantial evidence of a lack of a threshold in PM effects (Vedal, 1997). At the same time, while the standard appears to give weight to both acute and chronic effects, the rationale for an annual average is poorly developed.

Health risk assessments routinely describe uncertain parameters and coefficients as probability distributions and uses stochastic/Monte Carlo simulation methods to propagate that uncertainty. A simple "back-of-the-envelope" model illustrates this point. Equation (1) provides a simple model of risk/mortality from exposure to PM$_{2.5}$.

$$Risk(\#of\ excess\ deaths) = \alpha * C * M * P \quad (1)$$

where α is the percent change in annual mortality due to an increase of 1 μg/m3 in the annual average ambient PM$_{2.5}$ concentration C, M is background mortality rate (excluding trauma), and P is the exposed population. For the purpose of this illustration let us assume that C, M and P are exactly known and α is uncertain. Pope et al. (2002) estimated α = 0.4% for a cohort of roughly 500,000 US subjects age 30 and over. The Harvard Six Cities study (Dockery et al., 1993) estimated α = 1.3% for a cohort of about 8000 US subjects age 25 and older. In this calculation, uncertainty arises from the uncertainty in α which can be modeled either parametrically as above or by using a probability distribution. These two studies can be used to estimate a range for the number of excess deaths at a given annual average PM$_{2.5}$ concentration (set in this illustration at 25 μg/m^3) to be between 17,200 and 50,700 per year. M and P are set to current values for the US population and α can also be treated as a random variable with a known probability distribution and a probability distribution of excess deaths can be calculated using Monte Carlo simulation. Similarly, the variability in some model parameters such as the differing annual average PM$_{2.5}$ concentration in different geographical locations, or measurement errors in characterizing mortality rates can similarly be treated in a probabilistic manner. Thus, despite substantial uncertainties in particle composition and toxicological mechanisms, epidemiological studies can be used calculate health risk outcomes when uncertainty is incorporated.

There are several interesting insights from the evolution of our knowledge of risks from PM$_{2.5}$.

Initially, there was significant disagreement about the epidemiological findings and questions about the causal attribution of the health effects to $PM_{2.5}$ remained (Lam et al., 2006). Over time, the uncertainties in the results of early epidemiological studies were reduced through the execution of larger studies. A number of studies focused on the physical aspects and our knowledge about chemical characteristics of $PM_{2.5}$ and thus insight into mechanisms of toxicity has increased, but is not complete. Significant uncertainty remains about mechanisms linking particle characteristics to toxicity and health impact mechanisms. Interestingly, despite these uncertainties epidemiological evidence has resulted in a scientific consensus that makes it possible to quantify $PM_{2.5}$ risks. In other words, uncertainties are deemed to be "manageable", and risk assessment models that provide probability distributions for $PM_{2.5}$ risks are routinely used.

Health risks from nanoparticles share some features of the $PM_{2.5}$ risk calculations because of their size and estimated tissue residence time. However, the relevant parameters of nanoparticles may be more complex making it more difficult to estimate than those for $PM_{2.5}$. At present little information is available regarding potential toxicity of nanoparticles (Lam et al., 2006). The fundamental mechanisms for health risks from nanoparticles are largely unknown, and unlike the $PM_{2.5}$ case, no epidemiological data are available (Wichmann et al., 2000). In order to gain a better estimation of underlying basis for toxicity of nanoparticles information regarding specificity, toxicokinetic parameters, localization and half-life, will be required.

Uncertainty in characterizing health risks from nanoparticles

Uncertainties in model parameters can result from variability within a study sample, inherent "randomness", and lack of scientific knowledge as discussed in the previous section. In many cases, the latter can translate into uncertainty about appropriate model forms. Model uncertainty is "frequently important when the system involved is sufficiently complex that key influences have not yet been identified" (Casman et al., 1999). In what follows we will argue that

model uncertainty is endemic to the problem of calculating nanoparticle risks. We shall show that doing risk calculations for nanoparticles leads to an explosion of model forms, rendering the uncertainty extreme.

In occupational environments with sources that produce engineered nanoparticles, the incidental nanoparticles created by multiple sources form a background. Since incidental nanoparticles have a variety of shapes, sizes, and compositions, assessing their risks is a complex task that needs to account for this natural variability as well as model uncertainty. However, if the intent is to assess the risks of engineered nanoparticles, the background particles can be subtracted. For the engineered nanoparticles, the source and particle characteristics are well-defined. While the natural variability is completely characterized, model uncertainties will still persist. However, in terms of a strategy for risk assessment, this is a useful first step. As for any chemical, model uncertainties in risk assessments of nanoparticles can be classified into three categories: those resulting from physical and chemical characterization of nanoparticles including the choice of an appropriate exposure metric; those resulting from uncertainty in dose and health end-points from different exposure routes; and those resulting from a lack of understanding of toxicity mechanisms.

The list of uncertainties for nanoparticles also holds true for PM2.5. However, as described in the previous section, risk assessment for PM2.5 has avoided dealing with these issues because of a plethora of epidemiological studies. Such studies don't currently exist for the case of nanoparticles. Future epidemiological studies should rightly be in occupational environments due to the higher concentrations expected in such environments. However, unlike PM2.5, they will be very specific to a particular type of engineered nanoparticle and not generalizable to other nanoparticles. Thus, the lesson of PM2.5 is that broadly generalizable epidemiological studies are not really possible for nanoparticles, and the only way forward is to grapple with the various sources of uncertainty

Particle characteristics

Size and agglomeration
Nanoparticles are smaller than 100 nm in diameter. Included in this definition are engineered

nanoparticles, unintentionally created nanoparticles, and ambient ultrafine particles. Physical and chemical characteristics are as varied as the processes that create the particles. While it is true that nanoparticles have been with us for millennia (e.g., as combustion byproducts), the risks from nanoparticles arise from the unknown properties of *novel* nanoparticles. Engineered nanoparticles may be structurally and compositionally homogeneous or heterogeneous or even be multi-functional (e.g., those developed for medical diagnostics and treatment). As particle size becomes smaller, a greater fraction of atoms are at the surface and quantum effects tend to increase surface reactivity. At the same time, nanoparticles have a tendency to agglomerate and form larger structures. Thus, agglomeration can lead to a reduction in the number of atoms at the surface with a reduction in surface energy. Since coagulation half-lives of nanoparticles are of the order of tens of microseconds to a few milliseconds (Preining, 1998), nanoparticle concentrations can decrease rapidly by agglomeration. Engineered nanoparticles, however, are specially coated to reduce agglomeration in order to exploit high surface reactivity for various useful ends. This increases the potential for human inhalation exposures to very small nanoparticles and also affects their disposition in the body and toxicity (Araujo et al., 1999; Kirchner et al., 2001; Oberdörster et al., 1994). Agglomeration properties of various engineered nanoparticles (e.g., single or multi-walled carbon nanotubes, nanoclay particles, zinc oxide nanoparticles, dendrimers, or fullerenes) are unknown, limiting our ability to estimate the size distribution of the airborne nanoparticles, and thus their fate in the human body after inhalation.

Particle shape

Prior experience with asbestos and other fibrous aerosols indicates that the shape of the particles (i.e., their length and diameter) has a profound effect on toxicity. Smaller diameter fibers penetrate deeper into the respiratory tract, while longer fibers are cleared more slowly (Mossman et al., 1990; Royal Society, 2004). Engineered nanoparticles come in various shapes such as spheres (e.g., dendrimers), tubes (e.g., SWCNT and MWCNT), plates (e.g., nanoclay flakes), fullerenes, and needles. While it seems likely that particle shape will affect the deposition, fate, and toxicity of the

particles in the human body (Jia et al., 2005), few data about these effects are available.

Chemical composition

Chemical composition of the surface and the bulk of engineered nanoparticles will affect toxicity. As mentioned earlier, surface coatings that modify the agglomeration properties of nanoparticles will have biological effects (Oberdörster et al., 2005a; Warheit et al., 2005a). Experiments using fullerene soot with different impurities (e.g., metallic endohedral fullerene) indicate that the pulmonary toxicity response depends on the types of nanomaterials and their impurities (Quan & Chen, 2005).

Choice of exposure metric

Several studies have suggested that at similar mass concentrations, nanometer size particles are more harmful than micron size particles (Oberdörster et al., 1995; Seaton et al., 1995; Donaldson et al., 1996, 2000; Lison et al., 1997; Peters et al., 1997b; Donaldson, 1999; Brown et al., 2000, 2001; Cullen et al., 2000; Tran et al., 2000; Utell & Frampton, 2000; Renwick et al., 2001; Dick et al., 2003; MacNee & Donaldson, 2003). One possible explanation for this might be that since the number of particles and particle surface area per unit mass increases with decreasing particle size and pulmonary deposition increases with decreasing particle size, dose by particle number or surface area will increase as size decreases. Lison et al. (1997) and Tran et al. (2000) have demonstrated a close association between aerosol surface area and inflammatory response when using a range of chemically inert materials with low solubility. Oxidative stress has been highlighted in a number of studies as being a significant mechanism underlying an indicated increase in toxicity within ultrafine and high specific surface-area particles (Stone et al., 1998; Donaldson et al., 2000; Dick et al., 2003). At the same time, some preliminary studies seem to indicate that in some cases, exposures to nanoparticles may be less inflammatory than micro-scale particles (Warheit et al., 2005b). Thus, there is considerable uncertainty about which metric of exposure – mass, surface area, or number concentration – is most relevant to determining health effects, although new screening procedures are being proposed (Oberdörster et al., 2005c).

Currently, there are very few measurements of nanoparticle exposures either in the workplace or in residential or ambient environments. Maynard et al. (2004) measured exposures to manufactured carbon nanotubes in a model workplace and found low mass concentrations with particles that were mainly in aggregate form. However, this does not always imply low counts for other metrics since exposure metrics do not always track each other, i.e., if one metric is high, the other metrics are not necessarily also high. Kuhlbusch et al. (2004) used a strategy of subtracting ambient concentrations of ultrafine particles from concentrations of nanoparticles in carbon black packaging operations. This strategy allows a characterization of the manufactured particles as distinct from those released unintentionally. In a study of exposures to unintended nanoparticles, Ramachandran et al. (2005) evaluated the exposures to diesel exhaust particulates of three occupational groups – bus drivers, parking garage attendants, and bus mechanics – using the mass concentration of elemental carbon (EC), surface area, and number concentrations as the exposure metric. Nearly all of the mass emitted by engines is in the fine particle range (between 100 nm and 1 µm) and nearly all the number is in the nanoparticle range. The exposed three groups had significantly different exposures to workshift EC with the highest levels observed in the bus garage mechanics and the lowest levels in the parking ramp booth attendants. In terms of surface area, parking ramp attendants had significantly greater exposures than bus garage mechanics, who in turn had significantly greater exposures than bus drivers. In terms of number concentrations, the exposures of garage mechanics exceeded those of ramp booth attendants by a factor of 5–6. The three occupational groups had quite different exposure rankings for the three metrics, illustrating the importance of the choice of exposure metric in exposure and epidemiological studies. The exposure rankings of the three groups also change with the metric used. If the incorrect metric were to be used, significant misclassification errors could occur. The choice of an exposure metric is further complicated by physical transformation of nanoparticles after their release to the environment. Nanoparticles may or may not agglomerate to form larger particles and can also change shape adding an additional layer of uncertainty in characterizing atmospheric mass, surface area or number concentration.

From the perspective of risk assessment the uncertainty over exposure metric could lead to two eventual outcomes. The first is that a consensus will emerge on a particular metric that will vastly simplify the current situation. Alternatively, an exposure metric might be dependent on the type of nanoparticle, and be determined by such factors as chemical composition and particle morphology. The latter will vastly complicate the risk assessment process. Until such time as the issue of exposure metric is resolved, risk calculations face a fundamental model uncertainty, i.e., one of choosing the right variable in a model.

Exposure route

Inhalation
Particles deposit in different regions of the respiratory tract depending on their size and the particles' inertial, gravitational, and diffusional behavior (ICRP, 1994). For particles above 100 nm, the predictions of the ICRP model have been experimentally validated by a number of studies (reviewed in James et al., 1994; US EPA, 1996; Lippmann, 1999; Phalen, 1999). For nanoparticles, there are a few studies for particle deposition in the tracheobronchial and alveolar region (Heyder et al., 1986; Jaques & Kim, 2000) and extrathoracic (nasal) region (Cheng et al., 1996) that show reasonable agreement with ICRP model predictions. For nanoparticles, diffusion is the main mechanism for deposition and the deposition efficiency (number of particles of a given size deposited in a given region as a fraction of the number of particles entering that region) is highest in the alveolar region, followed by the tracheobronchial and extrathoracic regions (the same relationship holds for surface area and mass of particles). Most of the mass is deposited in the alveolar region (with progressively smaller amounts deposited in the tracheobronchial and nasal regions). In terms of dose to individual cells, the mass deposited may have to be normalized by the surface area of the epithelial cells, in which case the nasal and upper tracheobronchial regions receive much higher doses than the alveolar region (Oberdörster et al., 2005a). This dose may be more relevant to nanoparticle translocation to other sites than to local toxicological effects.

Dermal

Despite the fact that a number of commercial products containing nanoparticles such as cosmetics and suntan lotions are already available, there is considerable uncertainty about whether this is a significant exposure route. Most existing studies appear to indicate that few nanoparticles penetrate through the stratum corneum and the dermis (Tan et al., 1996; Lademan et al., 1999; SCCNFP, 2000; Schulz et al., 2002). Researchers also caution that leaching of selected components of the particles through the skin and into the blood stream is possible. Confounding these limited findings is the fact that different studies used different experimental protocols, making cross-study comparisons difficult.

Ingestion

Drug development based on engineered nanoparticles is one of nanotechnology's great promises, and oral ingestion is likely to be an important exposure route. Several studies have already shown that nanoparticles are efficiently absorbed through the gastrointestinal tract (Jani et al., 1992, 1994) and that the particles then translocate through the mucosal tissue into the lymphatic and circulatory systems (Moghimi et al., 2001). The risk from accidental exposures to nanoparticles via this route is uncertain, however.

Translocation and health effect endpoints

A critical difference between nanoparticles and particles of larger size is the ability to nanoparticles to move or 'translocate' to different parts of the body. There is evidence for the translocation of inhaled nanoparticles from the alveolar spaces into the interstitium (Oberdörster et al., 2005a), to local and regional lymph nodes (Oberdörster et al., 1994), and into the circulatory system (e.g., Nemmar et al., 2002; Kreyling et al., 2002). Translocation into the blood stream may also affect the ability of blood to coagulate (Nemmar et al., 2002). There is considerable uncertainty regarding the extent of translocation, which may depend on the surface and chemical characteristics of the particles. More recently, a growing body of evidence suggests that nanoparticles may translocate along the neuronal pathways into the central nervous system and the brain (Oberdörster et al., 2005b). However, the exact mechanisms of transport have not been elucidated and there may yet be other pathways in addition to neuronal transport. The small size facilitates uptake into cells and transport across epithelial and endothelial cells into the blood and lymph circulation, which may carry nanoparticles to potentially sensitive target sites such as bone marrow, lymph nodes, spleen, and heart. The ability to reach new regions, unhindered by natural blocking mechanisms of the body is, not surprisingly, the reason that nanoparticles may be useful in medical applications.

The fate and transport of larger particles is well-understood and inhalation is the exposure route. The preceding discussion shows that for nanoparticles, several exposures routes are possible. This is summarized in Figure 2, which illustrates that there are several pathways and potential end points for assessing the adverse health effects of nanoparticles in the body. In the case of inhalation, larger particulates are deposited in the lungs and can cause localized health effects within the pulmonary system. Nanoparticles, on the other hand, could have multiple health endpoints. For example, inhaled nanoparticles might translocate to the central nervous system either through neural pathways in the olfactory system or through the body's circulatory systems (blood and lymphatic) after absorption in the lungs and consequent crossing of the blood–brain barrier. The ability of nanoparticles to translocate to parts of the body that are not accessible to larger particles in the PM range or to most toxic chemicals and carcinogens makes it harder to extrapolate known laboratory and epidemiological findings to nanoparticles. In other words, existing physiological models of exposure and movement of particles within the human body are not sufficient.

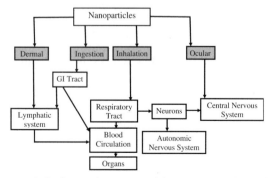

Figure 2. Pathways for nanoparticle exposure and translocation.

Toxicity mechanisms

Mechanisms of toxicity for nanoparticles have much in common with those for PM2.5. But the differences are sufficient to prevent easy extrapolation from our knowledge of the toxicity of larger particles (which is also marked with significant uncertainties). There are several mechanisms with varying degrees of evidence in their support.

Epidemiological studies have reported adverse human health effects that include respiratory and cardiovascular effects (e.g., Peters et al., 1997b; Wichmann et al., 2000; Penttinen et al., 2001). Controlled human exposure studies have reported effects on blood coagulation and systemic inflammation (e.g., Brown et al., 2002; Chalupa et al., 2004; Wichmann et al., 2000). There is a general understanding that changes in cell signalling pathways and gene expression as a result of oxidative stress might be the primary causes of the observed health effects. Several possible mechanisms may be able to accomplish this: (a) particle surface area (irrespective of chemical composition) can cause oxidative stress; (b) transition metals can cause the oxidative stress leading to intracellular calcium and gene activation; (c) cell surface receptors can be activated by transition metals which in turn cause gene activation; and (d) intracellular transport of nanoparticles to the mitochondria could lead to oxidative stress.

The shape of the nanoparticles appears to have an effect on toxicity: single walled carbon nanotubes are more toxic than multi-walled nanotubes, which are in turn more toxic than C_{60} fullerenes (Oberdörster et al., 2005a). It is still unclear whether knowledge from fiber toxicology is applicable to nanotubes. Further, nanoparticles tend to translocate to other regions of the body with great ease. The local toxic effects at relocated sites are yet to be studied in any depth. Particle size is an important determinant of toxicity. Smaller particles induce a greater inflammatory response and have a more pronounced effect on clearance mechanisms. There are several hypotheses to explain the size dependence of toxicity: (a) particle deposition characteristics; (b) greater specific surface area for smaller particles that leads to a disruption of chemotactic signals and impaired clearance of alveolar macrophages; and (c) surface characteristics. All these hypotheses need to be further investigated.

Expert judgment in uncertainty assessment

The previous section illustrated the deep uncertainties that pervade every element of the exposure–response–risk paradigm for nanoparticles. Risk assessment frequently needs to be performed in the absence of relevant measurement data or deterministic models. In such cases, professional or expert judgment can be useful in assessing relevant variables, characterizing model forms and estimating model parameters. Subjective or Bayesian methods for handling uncertainty have a long history originating with the use of Delphi method in technology forecasting and nuclear deterrence (Helmer, 1966; Kahn & Wiener, 1967; Linstone & Turoff, 1975), with subsequent applications in policy analysis, engineering, and risk analysis (Morgan & Henrion, 1990).

Most often, in scientific domains expert judgment is used to quantify uncertain parameters in a probabilistic form. However, expert judgment is not solely limited to assessing model parameters. Often, and especially in early stages of a scientific issue when uncertainty is high, expert assessment is used to structure problems, to indicate key variables, and to examine relationships and influences between variables by building 'influence diagrams' (Morgan & Henrion 1990). Here, the attempt is to develop a consistent representation as a way to structure scientific understanding[1] without necessarily quantifying parametric values or the strength of influences. This is best derived using qualitative descriptions of phenomena obtained through expert interviews. Further, even when expert judgment tasks are focused on eliciting parametric values, the task typically entails some amount of qualitative exploration of the phenomena at hand (such as descriptions of assumptions, reasons for including or excluding certain variables) before quantitative values are elicited.

Expert judgment approaches have been used in environmental applications such as environmental exposure assessment (Hawkins & Evans, 1989; Ramachandran & Vincent, 1999; Ramachandran,

---

[1] Some Bayesian approaches such as Bayesian Belief Networks also quantify influences between variables in a probabilistic manner. In doing so, they attempt both to give structure to the models, and also assess the impact of parametric and other uncertainties on model outputs.

148

2001; Walker et al., 2001, 2003; Ramachandran et al., 2003), and assessment of global climate change (Morgan & Keith, 1995; Risbey et al., 2000; Risbey & Kandlikar, 2002). In a series of papers, Walker et al. (2001, 2003) evaluated the use of expert judgment to estimate non-occupational exposures to benzene. The authors brought the experts together into a workshop where the aims of the project and the existing literature regarding benzene in the ambient environment were discussed. Selected papers had previously been provided to the experts. Later, each expert was interviewed to directly obtain the estimates of exposure with little or no decomposition of the judgment into its constitutive sub-parts. In contrast, Ramachandran and Vincent (1999), Ramachandran (2001), and Ramachandran et al. (2003) used expert judgments to elicit specific parameters that served as inputs to a predictive exposure model.

Recently Morgan (2005) used the expert assessment approach to develop influence diagrams to assess human health risks from nanoparticles. In this work, hierarchically nested 'modules' of influences were inferred for each element in the risk assessment causal chain using expert interview material. In other words, the project implicitly combined the understanding of different experts into a single influence diagram representing a form of collective understanding, an "uberexpert" if you will. Influence diagrams are very useful devices for structuring problems and can be used quantitatively if sufficient data are available about the quantitative relationships among variables. However, in the absence of such data, they are more useful as qualitative tools and cannot be used for risk assessment. In the next section, we suggest a simple approach that helps us start in that direction.

Quantifying uncertainty in nanoparticle risks: a modest proposal

The central challenge in quantifying nanoparticle risks is the presence of deep uncertainty as manifested in difficulties of choosing appropriate model variables, and the presence of multiple and competing models. Some of this uncertainty is irreducible in that there is ignorance about causal mechanisms that could influence risk calculations. One way to make these problems analytically tractable is to study the degree of consensus and/or

disagreement between experts on different parts of the exposure–response paradigm (Figure 3). This can be done by eliciting judgments from a wide range of experts on different parts of the risk causal chain, i.e., experts will be asked to assess uncertainties in source characteristics of nanoparticles, in their fate and transport through the environment, human exposures, in the choice of the proper exposure metric, in the modes of translocation within the human body, and in the mechanisms of toxicity.

The expert judgment process begins with a set of qualitative questions administered in an open-ended interview format to a small group of experts. The procedure of open-ended interviews with a small number of experts is common in many scientific expert assessment endeavors. These interviews lay the groundwork for developing expert assessment protocol to be administered to a larger group of experts. We call this a protocol rather than a survey because of the effort to provide structure and detail in setting out the relevant questions. It is not a survey of overall opinions on nanoparticle risks, but a tool to identify judgments and provide diagnostic guidance. The expert assessment protocol will require experts to independently provide answers to each question in a probabilistic fashion. The term probabilistic is used in this case to refer to a either a "degree of belief" or a subjective probability distribution of a variable. Questions could be asked in the following categories:

Subjective degrees of belief/agreement with specific statements
Experts will respond to 10 specific statements in all the four domains. The extent of their agreement with statement will be noted using the Likert scale (Strongly Agree – Agree – Undecided – Disagree or Strongly Disagree). The statements will be posed as definitive propositions. Two example statements are provided below:

a. The smaller the particle, the more harmful it is. This statement applies to all substances.
b. Manufactured nanoparticles (e.g., fullerenes, carbon nanotubes) are more likely to persist in the environment than other nanoparticles (e.g., generated as a byproduct combustion) due to different agglomeration processes resulting from special coatings on the particles.

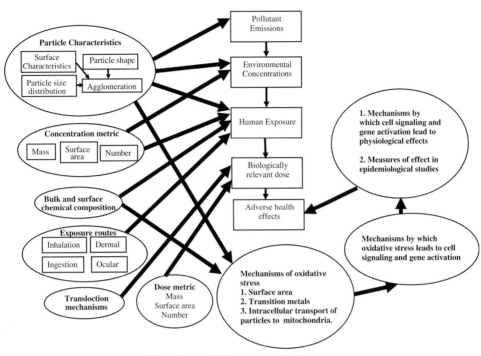

Figure 3. Nanoparticle risk framework with major variables and mechanisms.

Response to such statements helps us develop a broad overview of expert opinions on nanoparticle health risks. Further, since such questions can be asked of all experts independent of the domain of expertise, they are useful in evaluating the extent to which an experts' area of expertise correlates (or not) with specific views of nanoparticle risks.

Relative importance of variables
Experts could be asked to evaluate the importance of specific variables using ranking exercises or by assigning subjective degrees of belief ("probability"). *For example, in a question focusing on choice of exposure metric – mass or number of surface area concentration, the expert will be asked to assign points between 0 and 100 (corresponding to a probability measure) to each of the choices so that they add up to 100. The experts could be advised to award more points to the choice they consider more likely.* In this way a subjective probability measure that reflects the expert's state of belief regarding the choice of exposure metric can be obtained. A hypothetical response to this question may have the expert assigning a 30 points (probabil-

ity = 0.3) to mass concentration as the appropriate exposure metric, 60 points (probability = 0.6) to surface area concentration as the appropriate exposure metric and 10 points (probability = 0.1) that particle number is the appropriate exposure metric. Some variables may require more 'levels' of probability elicitation. For example, a set of questions could interrogate the expert's belief about the role of particle shape as a determinant of toxicity. The first level could ask if the expert considers shape to be highly important, moderately important or of negligible importance. The expert would assign probability scores to each of these three options that would add up to 100%. If the expert assigns a high probability to the option that shape is of negligible importance, then no further questions about particle shape are required. If the expert regards shape as an important determinant, then the next layer of questions will evaluate her beliefs about the importance of different shapes (e.g., cylindrical vs. fullerene) and then further branch into questions of how shape might affect toxicity.

This procedure can be implemented for each element in Figure 3 and will result in the assignation of probabilistic information to each element of the causal chain. Eliciting opinions from a large group of experts (20 or more) would assure that the causal chains [with the associated probabilities] would be sufficiently populated so as to allow meaningful comparisons across expert responses. On some elements of the causal chain, the experts may have very divergent opinions, whereas on other elements there may be a general consensus as reflected by the assigned probabilities. Figure 4 shows a hypothetical pictorial representation of how consensus and divergence in different elements of the causal chain can be evaluated. Here, each parameter in the causal chain is plotted along the vertical axis. The horizontal axis shows the level of an expert's belief in the importance of that parameter, and this can range from 0 to 1. In this figure, six expert opinions are shown, each with its own symbol. There appears to be general agreement about Parameters 1 and 2 (although one expert differs from all others). There is unanimity about Parameter 3 and 4, and there is broad disagreement about Parameters 5 and 6. Opinion is evenly divided about Parameter N. Figure 4 is a hypothetical example used only for the purpose of illustrating the concept, and the actual questions and the resulting causal chain might be more complex.

Relative importance of competing causal mechanisms to characterize model uncertainty

In these tasks experts could be provided with influence diagrams of competing hypothesis for causal mechanisms. Each hypothesized mechanism would have a corresponding influence diagram. Experts may modify, remove from or add to these diagrams. As in the example described above, experts would use a fixed 100-point scale to allocate their degree of belief ("probability" of being correct) that a mechanism represented by an influence diagram is the correct one.

Assigning probability values to variables

Case studies of nanoparticles could be used to assess whether there is sufficient knowledge to allow for calculation of health risks. Nanoparticle types would be chosen based on characteristics such as morphology, chemical composition, and surface coatings. A simple dose–response model where each step of the causal chain is characterized by small number of parameters would be developed, and experts would be asked to provide quantitative probability assessments in the form of PDFs for each of these parameters. Since uncertainties in these parameters are likely to be extreme not all experts may be willing/able to provide PDFs. Expert could choose chose a range of metrics in making these assessments – from full

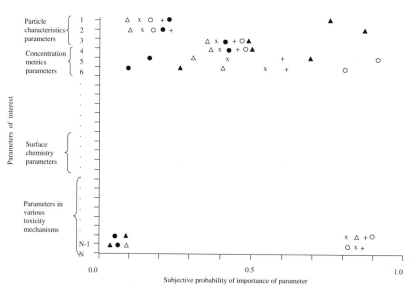

Figure 4. Hypothetical representation of consensus and divergence among experts about model parameters.

blown PDFs, to extreme bounds, order of magnitude assessment, and complete ignorance using the approach outlined in Kandlikar et al. (2005). The intent of such an exercise is not to actually perform a credible health risk calculation but to understand the gap between current expertise and the needs of a credible formal risk assessment.

Qualitative responses

In expert elicitation tasks, experts are asked to back up their quantitative responses with qualitative reasoning. Qualitative data is particularly helpful in understanding underlying reasons for why experts make the quantitative judgments they do. In addition, this procedure is typically adopted to combat overconfidence, a common trait observed in experts (Morgan & Henrion, 1990). Qualitative responses could be transcribed from the expert interviews, and classified into a small number of categories and tagged to the quantitative judgments. In this way it would become possible to observe whether experts use similar lines of reasoning when they provide closely matched quantitative responses to questions and vice-versa

Levels of expert consensus are a useful way to understand where the uncertainties are highest but care must be taken in interpreting the results. For instance, experts might overwhelmingly agree on a particular metric or mechanism as being the 'correct' one, in which case resources might be allocated to other topics. On the other hand, expert consensus may be based on limited data and experts may differ on the underlying reasons even when the results of quantitative assessment show agreement. Conversely, experts may be polarized and disagree strongly on whether a particular metric is appropriate suggesting that resources might need to be devoted to this specific question.

The derivation of consensus among the panel members can be approached from at least two angles: the behavioral and the mathematical. In general, behavioral approaches rely on psychological factors and interactions among experts, and mathematical schemes use a designated functional aggregation rule that accepts inputs from each expert and returns an arbitrated consensus (Winkler, 1968, 1986; Genest & Zidek, 1986). Behaviorally derived agreements often suffer from problems of personality and group dynamics. Mathematical approaches avoid these problems but introduce their own set; numerically dictated

compromises may be universally unsatisfactory. The last option is to not seek consensus at all, but to use each individual probabilistic judgment separately. This has the advantage of making explicit the differences in judgments.

Expert assessment raises some important methodological issues and there is large literature on this topic (reviewed in Morgan & Henrion, 1990; Cooke, 1991). The individuals selected as experts should meet several well-established criteria to qualify as "experts". These include substantive contributions to the scientific literature (Wolff et al., 1990), status in the scientific community, membership on editorial committees of key journals (Evans et al., 1994; Siegel et al., 1990) advisory boards, and peer nomination (Hawkins & Evans, 1989). The assembled team should include experts from industry and academia, and thus provide for variety and balance of institutional perspectives. The quality of expert judgment depends on (a) substantive expertise, referring to the knowledge that the expert has about the subject or variable of interest, and (b) normative expertise, referring to the skill in expressing beliefs in probabilistic terms, also known as calibration. Substantive expertise can be assured to some extent by the expert selection process. Normative expertise or calibration is a measure of the accuracy of the expert judgment and initial training of the experts in probabilistic responses might be useful in calibrating them prior to eliciting their judgments.

Conclusions

In many situations, there is a lack of sufficient information about multiple components of this risk framework. In such cases, agencies such as the US EPA have resorted to the use of default assumptions and extrapolations to fill in the data gaps. In the case of nanoparticle health risks the uncertainties are so large as to obviate any simple extrapolations. Our brief review of the risk assessment for $PM_{2.5}$ yielded some useful insights about nanoparticle risks. Despite significant and continuing uncertainty about mechanisms linking particle characteristics to toxicity and health impact mechanisms, quantitative risk assessment for $PM_{2.5}$ has made progress. It may well be that $PM_{2.5}$ is merely a surrogate for the true, but as yet unknown, exposure metric. But it is a reasonably good surrogate that large-scale epidemiological studies have yielded remarkably consistent results.

Health risks from nanoparticles share some features of the $PM_{2.5}$ risk calculations, though they may be more difficult to characterize than those from $PM_{2.5}$. The fundamental mechanisms for health risks from nanoparticles are largely unknown and extreme uncertainty prevails in almost every aspect of the exposure–response paradigm. No epidemiological data are available for engineered nanoparticle exposures, and it is unlikely that such data will be available in the near future in the absence of scientific consensus on the proper exposure metric and relevant health effect to be measured. Thus, while comprehensive risk assessment using is far from feasible, even simple screening assessments are difficult due to the novelty of the materials and the lack of basic toxicity data. Meanwhile, the technology is moving at a rapid pace, new nanoparticles are routinely manufactured and the social costs of 'getting it wrong' are very high. In such a situation, there is a need for quantitative frameworks that help in risk assessment and expert judgment may be a valuable tool for doing this.

We realize that doing a comparative study of expert opinions and examining the extent of expert consensus and disagreement can take us only so far. The point of this exercise is not to attempt to resolve the uncertainties regarding the health effects of nanoparticles by arriving at some artificial consensus among experts. All experts could be wrong on a particular issue, in which case consensus would be meaningless. Further, even a perfectly executed expert assessment exercise cannot tell us what we do not already know about the science. Rather, the goal is to provide a quantitative synthesis of the state-of-the-art where, in contrast with assessment done by committees, viewpoints of individual scientists can be highlighted and differences between scientists brought to fore. It is possible that an initial exercise of the type proposed here might reveal a complete lack of consensus among experts, and might not aid in risk assessment at all. However, this type of expert judgment elicitation can be easily repeated over time as collective knowledge evolves and uncertainties narrow leading to refined risk estimates. Such data can be useful in setting priorities, and allocating budgets in a manner that incorporates

diversity of scientific opinion. In the emerging and meta-stable science of nanotechnology, risk assessment keeping track of the diversity of scientific opinion could be very useful.

References

Anderson E.L. & D.H. Hattis, 1999. When and how can you specify a probability distribution when you don't know much? Risk Anal. 19, 43–68.

Araujo L., R. Lobenberg & J. Kreuter, 1999. Influence of the surfactant concentration on the body distribution of nanoparticles. J. Drug Target. 6, 373–385.

Barnes P.J., 2001. Neurogenic inflammation in the airways. Respir. Physiol. 125, 145–154.

Brown D.M., M.R. Wilson, W. MacNee, V. Stone & K. Donaldson, 2001. Size-dependent proinflammatory effects of ultrafine polystyrene particles: A role for surface area and oxidative stress in the enhanced activity of ultrafines. Toxicol. Appl. Pharmacol. 175(3), 191–199.

Brown D.M., V. Stone, P. Findlay, W. MacNee K. Donaldson, 2000. Increased inflammation and intracellular calcium caused by ultrafine carbon black is independent of transition metals or other soluble components. Occup. Environ. Med. 57(10), 685–691.

Brown J.S., K.L. Zeman & W.D. Bennett, 2002. Ultrafine particle deposition and clearance in the healthy and obstructed lung. Am. J. Respir. Crit. Care Med. 166, 1240–1247.

Burnett R.T., J. Brook, T. Dann, C. Delocla, O. Philips, S. Cakmak, R. Vincent, M.S. Goldberg & D. Kreski, 2000. Association between particulate- and gas-phase components of urban air pollution and daily mortality in eight Canadian cities. Inhal. Toxicol. 12(Suppl. 4), 15–39.

Campen M.J., J.P. Nolan, M.C.J. Schladweiler, U.P. Kodavanti, P.A. Evansky, D.L. Costa & W.P. Watkinson, 2001. Cardiovascular and thermoregulatory effects of inhaled PM-associated transition metals: A potential interaction between nickel and vanadium sulfate. Toxicol. Sci. 64, 243–252.

Campen M.J., W.P. Watkinson, J.R. Lehmann & D.L. Costa, 1996. Modulation of residual oil fly ash (ROFA) particle toxicity in rats by pulmonary hypertension and ambient temperature. Am. J. Respir. Crit. Care Med. 153, A542.

Casman E.A., M.G. Morgan & H. Dowlatabadi, 1999. Mixed levels of uncertainty in complex policy models. Risk Anal. 19(1), 33–42.

Chalupa D.C., P.E. Morrow, G. Oberdörster, M.J. Utell & M.W. Frampton, 2004. Ultrafine particle deposition in subjects with asthma. Environ. Health Perspect 112, 879–882.

Cheng Y.S., H.C. Yeh, R.A. Guilmette, S.Q. Simpson, K.H. Cheng & D.L. Swift, 1996. Nasal deposition of ultrafine particles in human volunteers and its relationship to airway geometry. Aerosol Sci. Technol. 25, 274–291.

Coleman K.P., W.A. Toscano Jr. & T.E. Wiese, 2003. QSAR models of the in vitro estrogen activity of bis phenol A analogs. QSAR Comb. Sci. 22, 78–88.

Cooke R.M., 1991. Experts in Uncertainty: Opinion and Subjective Probability in Science. Oxford University Press.

Costa D.L., J.R. Lehmann, L.T. Frazier, D. Doerfler & A. Ghio, 1996. Pulmonary hypertension: A possible risk factor in particulate toxicity. Am. J. Respir. Crit. Care Med. 149, A840.

Cullen R.T., C.L. Tran, D. Buchanan, J.M.C. Davis, A. Searl, A.D. Jones & K. Donaldson, 2000. Inhalation of poorly soluble particles. I. Differences in inflammatory response and clearance during exposure. Inhal. Toxicol. 12(12), 1089–1111.

Dick C.A.J., D.M. Brown, K. Donaldson & V. Stone, 2003. The role of free radicals in the toxic and inflammatory effects of four different ultrafine particle types. Inhal. Toxicol. 15(1), 39–52.

Dockery D.W., C.A. Pope III, X. Xu, J.D. Spengler, J.H. Ware, M.E. Fay, B.G. Ferris Jr. & F.E. Speizer, 1993. An association between air pollution and mortality in six U.S. cities. N. Engl. J. Med. 329, 1753–1759.

Donaldson K., 1999. In: Shuker L. and Levy L. eds. Mechanisms for Toxicity: In vitro; IEH Report on: Approaches to Predicting Toxicity from Occupational Exposure to Dusts. Report R11. Page Bros., Norwich, UK, pp. 17–26.

Donaldson K., P.H. Beswick & P.S. Gilmour, 1996. Free radical activity associated with the surface of unifying factor in determining biological activity? Toxicol. Lett.(Pg) 1–3.

Donaldson K., V. Stone, P.S. Gilmore, D.M. Brown & W. MacNee, 2000. Ultrafine particles: Mechanisms of lung injury. Phil. Trans. R. Soc. Lond. A 358, 2741–2749.

Dreher K., R. Jaskot, J. Richards & J. Lehmann, 1996. Acute pulmonary toxicity of size-fractionated ambient air particulate matter. Am. J. Respir. Crit. Care Med. 153, A15.

Evans J.S., G.M. Gray, R.L. Sielken, A.E. Smith, C. Valdez-Flores & J.D. Graham, 1994. Use of probabilistic expert judgment in distributional analysis of carcinogenic potency. Risk Anal. 20, 15–34.

Fairley D., 1999. Daily mortality and air pollution in Santa Clara County, California: 1989–1996. Environ. Health Perspect. 107, 637–641.

Ferin J., G. Oberdörster & D.P. Penney, 1992. Pulmonary retention of ultrafine and fine particles in rats. Am. J. Respir. Cell Molec. Biol. 6, 535–542.

Genest C. & J.V. Zidek, 1986. Combining probability distributions: A critique and an annotated bibliography. Stat. Sci. 1, 114–148.

Ghio A.J. & R.B. Devlin, 2001. Inflammatory lung injury after bronchial instillation of air pollution particles. Am. J. Respir. Crit. Care Med. 164, 704–708.

Ghio A.J., C. Kim & R.B. Devlin, 2000. Concentrated ambient air particles induce mild pulmonary inflammation in healthy human volunteers. Am. J. Respir. Crit. Care Med. 162, 981–988.

Godleski J.J., C. Sioutas, M. Katler & P. Koutrakis, 1996. Death from inhalation of concentrated ambient air particles in animal models of pulmonary disease. Am. J. Respir. Crit. Care Med. 153, A15.

Hahn F.F., E.B. Barr, M. Ménache & J.C. Seagrave, 2005. Particle Size and Composition Related to Adverse Health Effects in Aged, Sensitive Rats. Research Report 129, Health Effects Institute, Cambridge, MA.

Hawkins N.C. & J.S. Evans, 1989. Subjective estimation of toluene exposures: A calibration study of industrial hygienists. Appl. Ind. Hyg. 4, 61–68.

Helmer O., 1966 Social Technology. New York: Basic Book.

Heyder J., J. Gebhart, G. Rudolf, C.F. Schiller & W. Stahlhofen, 1986. Deposition of particles in the human respiratory tract in the size range 0.005–15 μm. J. Aerosol Sci. 17, 811–825.

Ibald-Mulli A., J. Stieber, H.-E. Wichmann, W. Koenig & A. Peters, 2001. Effects of air pollution on blood pressure: A population based approach. Am. J. Public Health 91, 571–577.

International Commission on Radiological Protection (ICRP), 1994. Human Respiratory Tract Model for Radiological Protection, ICRP Publication 66. Pergamon Press, Elmsford, NY.

James A.C., W. Stahlhofen, G. Rudolf, R. Köbrich, J.K. Briant, M.J. Egan, W. Nixon & A. Birchall, 1994. In: Deposition of Inhaled Particles; Human Respiratory Tract Model for Radiological Protection, *Annex D*, Annals of the ICRP. Pergamon Press, Oxford, UK, pp. 231–299.

Jani P.U., A.T. Florence & D.E. McCarthy, 1992. Further histological evidence of the gastrointestinal absorption of polystyrene nanospheres in the rat. Int. J. Pharm. 84, 245–252.

Jani P.U., D.E. McCarthy & A.T. Florence, 1994. Titanium dioxide (rutile) particle uptake from the rat GI tract and translocation to systemic organs after oral administration. Int. J. Pharm. 105, 157–168.

Jaques P.A. & C.S. Kim, 2000. Measurement of total lung deposition of inhaled ultrafine particles in healthy men and women. Inhal. Toxicol. 12, 715–731.

Jia G., H. Wang, L. Yan, X. Wang, R. Pei, T. Yan, et al., 2005. Cytotoxicity of carbon nanomaterials: Single wall nanotube, multiwall nanotube, and fullerene. Environ. Sci. Technol. 39, 1378–1383.

Kahn H. & A.J. Wiener, 1967. The Year 2000, A Framework for Speculation. New York: Macmillan.

Kandlikar M., J. Risbey & S. Dessai, 2005. Representing and communicating deep uncertainty in climate change assessment. C R Geosci 2337, 443–455.

Kirchner C., T. Liedl, S. Kudera, T. Pellegrino, J.A. Munoz, H.E. Gaub, et al., 2001. Cytotoxicity of colloidal CdSe and CdSe/ZnS nanoparticles. Nano Lett. 5, 331–338.

Kreyling W., M. Semmler, F. Erbe, P. Mayer, S. Takenaka, H. Schulz, et al., 2002. Translocation of ultrafine insoluble iridium particles from lung epithelium to extrapulmonary organs is size-dependent but very low. J. Toxicol. Environ. Health 65A, 1513–1530.

Kuhlbusch T.A., S. Neumann & H. Fissan, 2004. Number size distribution, mass concentration, and particle composition of PM1, PM2.5, and PM10 in bag filling areas of carbon black production. J. Occup. Environ. Hyg. 1(10), 660–671.

Lademann J., H.-J. Weigmann, C. Rickmeyer, H. Barthelmes, H. Schaefer, G. Mueller & W. Sterry, 1999. Penetration of titanium dioxide microparticles in a sunscreen formulation

154

into the horny layer and the follicular orifice. Skin Pharmacol. Appl. Skin Physiol. 12, 247–256.

Lam C.W., J.T. James, R. McCluskey, S. Arepalli & R.L. Hunter, 2006. A review of carbon nanotube toxicity and assessment of potential occupational and environmental health risks. CRC Crit. Rev. Toxicol. 36, 189–217.

Lison D., C. Lardot, F. Huaux, G. Zanetti & B. Fubini, 1997. Influence of particle surface area on the toxicity of insoluble manganese dioxide dusts. Arch. Toxicol. 71(12), 725–729.

Linstone H.A. & M. Turoff, 1975. The Delphi Method, Techniques and Applications. Reading, MA: Addison Wesley.

Lippmann M., 1999. Sampling Criteria for Fine Fractions of Ambient Air. In: Vincent J.H. ed. Particle Size-Selective Sampling for Particulate Air Contaminants. American Conference of Governmental Industrial Hygienists (ACGIH), Cincinnati, OH, pp. 97–118.

Lippmann M., K. Ito, A. Nadas & R.T. Burnett, 2000. Association of particulate matter components with daily mortality and morbidity in urban populations. Research Report 95, Health Effects Institute, Cambridge, MA.

MacNee W. & K. Donaldson, 2003. Mechanism of lung injury caused by PM10 and ultrafine particles with special reference to COPD. Eur. Resp. J. 21, 47S–51S.

Maynard A.D., P.A. Baron, M. Foley, A.A. Shvedova, E.R. Kisin & V. Castranova, 2004. Exposure to carbon nanotube material: Aerosol release during the handling of unrefined single-walled carbon nanotube material. J. Toxicol. Environ. Health A 67, 87–107.

Moghimi S.M., A.C. Hunter & J.C. Murray, 2001. Long-circulating and target-specific nanoparticles: Theory to practice. Pharmacol. Rev. 53, 283–318.

Moolgavkar S.H. & E.G. Luebeck, 1996. A critical review of the evidence on particulate air pollution and mortality. Epidemiology 7, 420–428.

Moolgavkar S.H., E.G. Luebeck, T.A. Hall & E.L. Anderson, 1995. Air pollution and daily mortality in Philadelphia. Epidemiology 6, 476–484.

Morgan M.G. & M. Henrion, 1990. Uncertainty: A Guide to Dealing with Uncertainty in Quantitative Risk and Policy Analysis. Cambridge: University Press.

Morgan M.G. & D. Keith, 1995. Subjective judgments by climate experts. Environ. Sci. Technol. 29(10), 468–476.

Morgan K., 2005. Development of a preliminary framework for informing the risk analysis and risk management of nanoparticles. Risk Anal. 25(6), 1–15.

Mossman B.T., J. Bignon, M. Corn, A. Seaton & J.B.L. Gee, 1990. Asbestos: Scientific developments and implications for public policy. Science 247, 294–301.

Nemmar A., P.H.M. Hoet, B. Vanquickenborne, D. Dinsdale, M. Thomeer, M.F. Hoylaerts, H. Vanbilloen, L. Mortelmans & B. Nemery, 2002. Passage of inhaled particles into the blood circulation in humans. Circulation 105, 411–414.

NRC/NAS Committee on the Institutional Means for Assessment of Risks to Public Health, Risk Assessment in the Federal Government (The Redbook), 1983.

Oberdörster G., E. Oberdörster & J. Oberdörster, 2005. Invited review: Nanotechnology: An emerging discipline evolving from studies of ultrafine particles. Environ. Health Perspect. 113(7), 823–839.

Oberdörster G., J. Ferin & B.E. Lehnert, 1994. Correlation between particle size, in vivo particle persistence, and lung injury. Environ. Health Perspect. 102(Suppl. 5), 173–179.

Oberdörster G., 2000. Toxicology of ultrafine particles: In vivo studies. Philos. Trans. R. Soc. Lond., Ser. A 358, 2719–2740.

Oberdörster G., R.M. Gelein, J. Ferin & B. Weiss, 1995. of particulate air pollution and acute mortality: Involvement of ultrafine particles? Inhal. Toxicol. 7, 111–124.

Oberdörster G., Z. Sharp, V. Atudorei, A. Elder, R. Gelein, W. Kreyling & C. Cox, 2005b. Translocation of inhaled ultrafine particles to the brain. Inhal. Toxicol. 16, 437–445.

Oberdörster G., A. Maynard, K. Donaldson, V. Castranova, J. Fitzpatrick, K. Ausman, J. Carter, B. Karn, W. Kreyling, D. Lai, S. Olin, N. Monteiro-Riviere, D. Warheit & H. YangILSI Research Foundation/Risk Science Institute Nanomaterial Toxicity Screening Working Group., 2005c. Principles for Characterizing the Potential Human Health Effects from Exposure to Nanomaterials: Elements of a Screening Strategy. Part. Fiber Toxicol. 2, 8–43.

Pekkanen J., E.J. Brunner, H.R. Anderson, P. Tiittanen & R.W. Atkinson, 2000. Daily concentrations of air pollution and plasma fibrinogen in London. Occup. Environ. Med. 57, 818–822.

Penttinen P., K.L. Timonen, P. Tiittanen, A. Mirme J. Ruuskanen & J. Pekkanen, 2001. Ultrafine particles in urban air and respiratory health among adult asthmatics. Eur. Resp. J. 17, 428–435.

Peters A., H.E. Wichmann, T. Tuch, J. Heinrich & J. Heyder, 1997b. Respiratory effects are associated with the number of ultra-fine particles. Am. Respir. Crit. Care Med. 155, 1376–1383.

Peters A., A. Doring, H.-E. Wichmann & W. Koenig, 1997a. Increased plasma viscosity during an air pollution episode: A link to mortality? Lancet 349, 1582–1587.

Peters A., M. Frohlich, A. Doring, T. Immervoll, H.-E. Wichmann, W.L. Hutchinson, M.B. Pepys & W. Koenig, 2001. Particulate air pollution is associated with an acute phase response in men; results from the MONICA-Augsburg Study. Eur. Heart J. 22, 1198–1204.

Phalen R.F., 1999. Airway Anatomy and Physiology. In: Vincent J.H. ed. Particle Size-Selective Sampling for Particulate Air Contaminants. American Conference of Governmental Industrial Hygienists (ACGIH), Cincinnati, OH, USA.

Pope C.A. III, R.T. Burnett, M.J. Thun, et al., 2002. Lung cancer, cardiopulmonary mortality, and long-term exposure to fine particulate air pollution. JAMA 287, 1132–1141.

Preining O., 1998. The physical nature of very, very small particles and its impact on their behaviour. J. Aerosol Sci. 29, 481–495.

Quan C. & L.C. Chen, 2005. In: Toxicity of Manufactured Nanomaterials; Proceedings of the 2nd International Symposium on Nanotechnology and Occupational Health, Minneapolis, MN, p. 24.

Ramachandran G., 2001. Retrospective exposure assessment using Bayesian methods. Ann. Occup. Hyg. 45(8), 651–667.

Ramachandran G. & J.H. Vincent, 1999. A Bayesian approach to retrospective exposure assessment. Appl. Occup. Environ. Hyg. 14, 547–557.

Ramachandran G., S. Banerjee & J.H. Vincent, 2003. Expert judgment and occupational hygiene: Application to aerosol speciation in the nickel primary production industry. Ann. Occup. Hyg. 47, 461–475.

Ramachandran G., W.F. Watts & D. Kittelson, 2005. Mass, surface area, and number metric in diesel occupational exposure assessment. J. Environ. Monit. 7(7), 728–735.

Renwick L.C., K. Donaldson & A. Clouter, 2001. Impairment of alveolar macrophage phagocytosis by ultrafine particles. Toxicol. Appl. Pharmacol. 172, 119–127.

Risbey J.S. & M. Kandlikar, 2002. Expert assessment of uncertainties in detection and attribution of climate change. Bull. Am. Meteorol. Soc. 1317–1326.

Risbey J.S., M. Kandlikar & D.J. Karoly, 2000. A protocol to articulate and quantify uncertainties in climate change detection and attribution. Climate Res. 16(1), 61–78.

Roco M.C., 2005, International perspective on government nanotechnology funding in 2005. JNR 7(6).

The Royal Society and The Royal Academy of Engineering, 2004. Nanoscience and Nanotechnologies: Opportunities and Uncertainties.

Salvi S., A. Blomberg, B. Rudell, F. Kelly, T. Sandstrom, S.T. Holgate & A. Frew, 1999. Acute inflammatory responses in the airways and peripheral blood after short-term exposure to diesel exhaust in healthy human volunteers. Am. J. Respir. Crit. Care Med. 159, 702–709.

Salvi S., C. Nordenhall, A. Blomberg, B. Rudell, J. Pourazer, F.J. Kelly, S. Wilson, T. Sandstrom, S.T. Holgate & A. Frew, 2000. Acute exposure to diesel exhaust increases IL-8 and GRO-a production in healthy human airways. Am. J. Respir. Crit. Care Med. 161, 550–557.

Samet J.M., S.L. Zeger & K. Berhane, 1995. In: The Association of Mortality and Particulate Air Pollution; Particulate Air Pollution and Daily Mortality: Replication and Validation of Selected Studies (The Phase I.A Report of the Particle Epidemiology Evaluation Project). Health Effects Institute, Cambridge, MA, pp. 3–104.

Samet J.M., S.L. Zeger, F. Domenici, F. Curreiro, I. Coursac, D.W. Dockery, J. Schwartz & A. Zanobetti, 2000. The National Morbidity, Mortality, and Air Pollution Study, Part II: Morbidity and Mortality from Air Pollution in the United States. Research Report 94, Health Effects Institute, Cambridge, MA.

Samet J.M., S.L. Zeger, J.E. Kelsall, J. Xu & L.S. Kalkstein, 1997. In: Weather, Air Pollution and Mortality in Philadelphia 1973–1980; Particulate Air Pollution and Daily Mortality: Analyses of the Effects of Weather and Multiple Air Pollutants (The Phae 1.B. Report of the Particle Epidemiology Evaluation Project). Health Effects Institute, Cambridge, MA, pp. 1–30.

Schulz J., H. Hohenberg, F. Pflücker, E. Gärtner, T. Will, S. Pfeiffer, R. Wepf, V. Wendel, H. Gers-Barlag & K.-P. Wittern, 2002. Distribution of sunscreens on skin. Adv. Drug Deliv. Rev. 54(Suppl. 1), S157–S163.

Schwartz J., 2001. Air pollution and blood markers of cartdiovascular risk. Environ. Health Perspect. 109(Suppl. 3), 405–409.

Schwartz J., 1994. Air pollution and daily mortality: A review and meta analysis. Environ. Res. 64, 36–52.

Schwartz J. & D.W. Dockery, 1992. Increased mortality in Philadelphia associated with daily air pollution concentrations. Am. Rev. Respir. Dis. 145, 600–604.

Schwartz J., D.W. Dockery & L.M. Neas, 1996. Is daily mortality associated specifically with fine particles. J. Air Waste Manage. Assoc. 46, 927–939.

Scientific Committee on Cosmetic and Non-Food Products (SCCNFP), 2000. Opinion concerning Titanium Dioxide (Colipa n S75). SCCNP: Brussels, 2000 www.europa.eu.int/comm/health/ph_risk/committees/sccp/docshtml/sccp_out135_en.htm.

Seaton A., A. Soutar, V. Crawford, R. Elton, S. McNerlan, J. Cherrie, M. Watt, R. Agius & R. Stout, 1999. Particulate air pollution and the blood. Thorax 54, 1027–1032.

Seaton A., W. MacNee, K. Donaldson & D. Godden, 1995. Particulate air pollution and acute health effects. Lancet 345, 176–178.

Service R.F., 2004. Nanotechnology grows up. Science 304, 1732–1734.

Sexton K., M.A. Callahan & E.F. Bryan, 1995. Estimating exposure and dose to characterize health risks: The role of human tissue monitoring in exposure assessment. Environ. Health Perspect. 103(Suppl. 3), 13–29.

Siegel J.E., J.D. Graham & M.A. Stoto, 1990. Allocating resources mong AIDS research strategies. Policy Sci. 23, 1–23.

Stone V., J. Shaw, D.M. Brown, W. MacNee, S.P. Faux & K. Donaldson, 1998. The role of oxidative stress in the prolonged inhibitory effect of ultrafine carbon black on epithelial cell function. Toxicol. In vitro 12(6), 649 (pp. 10).

Tan M.-H., C.A. Commens, L. Burnett & P.J. Snitch, 1996. A pilot study on the percutaneous absorption of microfine titanium dioxide from sunscreens. Aust. J. Dermatol. 37, 185–187.

Tran C.L., D. Buchanan, R.T. Cullen, A. Searl, A.D. Jones & K. Donaldson, 2000. Inhalation of poorly soluble particles. II. Influence of particle surface area on inflammation and clearance. Inhal. Toxicol. 12(12), 1113–1126.

US Environmental Protection Agency, 2004. An Examination of EPA Risk Assessment Principles and Practices. Office of the Science Advisor. EPA/100/B-04/001.

US Environmental Protection Agency, 1996. Air Quality Criteria for Particulate Matter, EPA/600/P-95/001cf.

Utell M.J. & M.W. Frampton, 2000. Acute health effects of ambient air pollution: The ultrafine particle hypothesis. J. Aerosol Med. 13(4), 355–359.

Vedal S., 1997. Ambient particles and health: Lines that divide. J. Air Waste Manage. Assoc. 475, 551–581.

Vincent R., P. Kumarathasan, P. Goegan, S.G. Bjarnason, J. Guenette, D. Berube, I.Y. Adamson, S. Desjardins, R.T. Burnett, F.J. Miller & B. Battistini, 2001. Inhalation Toxicology of Urban Ambient Particulate Matter: Acute Cardiovascular Effects in Rats. Research Report 104, Health Effects Institute, Boston, MA.

Walker K.D., P. Catalano, J.K. Hammitt & J.S. Evans, 2003. Use of expert judgment in exposure assessment: Part 2. Calibration of expert judgments about personal exposures to benzene. J. Expo. Anal. Environ. Epidemiol. 13, 1–16.

156

Walker K.D., D. Macintosh & J.S. Evans, 2001. Use of expert judgment in exposure assessment: Part I. Characterization of personal exposure to benzene. J. Expo. Anal. Environ. Epidemiol. 11, 308–322.

Warheit D.B., W.J. Brock, K.P. Lee, T.R. Webb & K.L. Reed, 2005a. Comparative pulmonary toxicity inhalation and instillation studies with different TiO_2 particle formulations: Impact of surface treatments on particle toxicity. Toxicol. Sci. 88(2), 514–524.

Warheit D.B., T.R. Webb, K.L. Reed, C. Sayes, Y. Liu & V.L. Colvin, 2005b. In: Pulmonary Effects of Nanoscale Titania and Quartz Particles: Role of Particle Size and Surface Area; Proceedings of the 2nd International Symposium on Nanotechnology and Occupational Health, Minneapolis, MN, p. 28.

Watkinson W.P., M.J. Campen & D.L. Costa, 1998. Cardiac arrhythmia induction after exposure to residual oil fly ash particles in a rodent model of pulmonary hypertension. Toxicol. Sci. 41, 209–216.

Wichmann H.-E., C. Spix, T. Tuch, G. Wolke, A. Peters, J. Heinrich, W.G. Kreyling & G. Heyder, 2000. Daily mortality and fine and ultrafine particles in Erfurt, Germany. Part I: Role of particle number and particle mass. Research Report 98, Health Effects Institute, Cambridge, MA.

Winkler R.L., 1986. Expert resolution. Manage. Sci. 32, 298–306.

Winkler R.L., 1968. The consensus of subjective probability distributions. Manage. Sci. 15(2), B61–B75.

Wolff S.K., N.C. Hawkins, S.M. Kennedy & J.D. Graham, 1990. Selecting experimental data for use in quantitative risk assessment: An expert judgment approach. Toxicol. Ind. Health 6, 275–295.

Journal of Nanoparticle Research (2007) 9:157–163
DOI 10.1007/s11051-006-9185-3

Special issue: Nanoparticles and Occupational Health

© Springer 2006

Evaluation of nanoparticle emission for TiO$_2$ nanopowder coating materials

Li-Yeh Hsu and Hung-Min Chein
*Energy and Environment Research Laboratories, Industry Technology Research Institute, Hsinchu, 310,
Taiwan, ROC (Tel.: +886-3-5913853; Fax: +886-3-5918753; E-mail: hmchein@itri.org.tw)*

Received 24 September 2006; accepted in revised form 2 October 2006

Key words: TiO$_2$, nanomaterials, nanoparticle emission, environment, occupational health

Abstract

In this study, nanoparticle emission of TiO$_2$ nanopowder coated on different substrates including wood, polymer, and tile, was evaluated in a simulation box and measured with a Scanning Mobility Particle Sizer (SMPS) for the first time. The coating process for the substrate followed the instructions given by the supply company. In the simulation box, UV light, a fan, and a rubber knife were used to simulate the sun light, wind, and human contacting conditions. Among the three selected substrates, tile coated with TiO$_2$ nanopowder was found to have the highest particle emission (22 #/cm^3 at 55 nm) due to nanopowder separation during the simulation process. The UV light was shown to increase the release of particle below 200 nm from TiO$_2$ nanopowder coating materials. The results show that, under the conditions of UV lamps, a fan and scraping motion, particle number concentration or average emission rate decreases significantly after 60 and 90 min for TiO$_2$/polymer and TiO$_2$/wood, respectively. However, the emission rate continued to increase after 2 h of testing for TiO$_2$/tile. It is suggested that nanoparticle emission evaluation is necessary for products with nanopowder coating.

Introduction

Due to their novel physical and chemical properties, nanomaterials have become one of the key components in the development of nanotechnology. The market value of the nanotechnology industry is estimated to be 1 trillion US dollars in 2015 (Roco, 2005). Major developed countries and international companies are funding many projects developing new materials and exploring their applications. However, the potential health risks and environmental impacts of nanomaterials are not understood and have raised public concerns. Nanomaterial manufacturing processes may potentially cause human exposure to nanomaterials via inhalation, dermal or ingestion routes (Aitken et al., 2004). In the past decade, health effects of nanoparticles (or ultra fine particles

(UFPs)) has been investigated by several research groups. Studies in animals using laboratory-generated or ambient UFPs show that they consistently induce mild yet noticeble pulmonary inflammatory responses as well as effects in extra pulmonary organs. Animal inhalation studies have included the use of different susceptibility models in rodents, with analysis of lung lavage parameters and lung histopathology, effects on the blood coagulation cascade and translocation studies to extra pulmonary tissues (Ferin et al., 1991; Ferin & Oberdörster, 1992; Oberdörster et al., 1992a, 1995, 2000, 2002, 2004; Li et al., 1999; Nemmar et al., 1999, 2002a, b, 2003; Elder et al., 2000, 2002, 2004; Kreyling et al., 2002; Zhou et al., 2003). Recently, principles for characterizing the potential human health effects from exposure to nanomaterials have been proposed by a group of distinguished scientists (Oberdörster et al., 2005). This work emphasized a screening strategy rather than a detailed testing protocol.

The exposure possibilities are apparent for researchers and workers in nanomaterial development and manufacturing. However, commercial products using or applying nanomaterials may increase the risk of exposure for consumers or end users. TiO_2 nanoparticles are being used increasingly in photocatalytic surface coatings. However, to our knowledge, there is no information in the open literature about possible emission of particle from nanomaterials which employ or apply nanopowder coating. Particle emission evaluation is critical to answer the question of whether such coatings will release nanoparticles during their use. This study aims to evaluate nanoparticle emissions from TiO_2 nanopowder coating materials in a simulated utilization stage. UV light/fluorescent lamps, a fan, and a rubber knife were used to simulate the sun light, wind, and human contact with the materials in a closed box.

Experiment

Materials

Anatase polycrystalline single crystal TiO_2 samples (Degussa Inc., German) were used as the coating agent. Wood plate and PET polymer film substrates were spray coated with 5 wt.% TiO_2 suspension and dried at 100°C for 2 h, designated as TiO_2/wood and TiO_2/polymer, respectively. In addition, a commercial TiO_2 photo catalyst paint (Allstar Painting Inc., Taiwan) was used as the second sample in this study. A tile plate was covered with a suitable amount of the TiO_2 paint and dried at room temperture. The final product was designated as TiO_2/tile. The above coating process

Figure 1. The simulation box. (a) appearance, (b) inner, and (c) the fan, rubber knife and substrate.

followed instructions from the paint supplier. The surfaces of these three substrates (wood plate, polymer film, and tile plate) were quite clean and smooth before the coating.

Simulation box

Sunlight, wind, and human contact with TiO_2 coating materials were simulated with UV/fluorescent lamps, a fan, and a rubber knife in a closed box as shown in Figure 1. The outside and interior of the simulation box are shown in Figure 1a, b, respectively. The light source was 5 UV lamps (wavelength = 365 nm, power density = 50 μW/cm^2) on the top inside the box. The fan and rubber knife were installed in the center of the simulation box and shown in Figure 1c. The rubber knife, simulating human contact, was operated with a motor and the fan was operating continuously during the test.

Before the emission measurement, 10 cm^2 of the substrate under test that had been coated with TiO_2 was cleaned using a clean air gun for 5 min and mounted on a small horizontal support at the bottom of the box. To obtain a stable background particle concentration, the box was purged with clean compressed air for several hours to make sure that particle generation inside the box was negligible (total concentration < 20 #/cm^3). The background particle concentration was deducted from the sample concentration. The operating conditions for the simulation experiments are listed in Table 1. The rubber knife scraping motion was performed for 1 min every 10 min, over 2 h. Particle emissions from the substrate and TiO_2 coating materials inside the box were monitored at different operation conditions. We repeated the 2-h particle emission measurement at least twice for each test condition. A Scanning Mobility Particle Sizer (SMPS, TSI Inc. USA) was used to measure the particle size distribution and number

Table 1. Operating conditions of three simulation experiments

Case	UV lamps 50 μW/cm^2)	Fluorescent lamps	Fan (75 m/min)	Rubber knife scraping motion
S1	√	X	√	√
2	√	X	√	X
3	X	√	√	√

Note: √ means operating. X means not operating.

concentration during tests. The SMPS is a combination of Electrostatic Classifier (TSI 3080L) and Ultra fine Condensation Particle Counter (UCPC, TSI 3025A). Particle size distributions ranging from 15 to 661 nm can be measured and displayed within the response time of 135 s (Chen & Chein, 2003).

Results and discussion

Particle emission measurement

Figure 2a shows the number concentration of the particles released from TiO_2/wood subjected to UV light, a fan and scraping motion (case 1 simulation) as a function of the time. In addition, Figure 2b shows the number concentration and

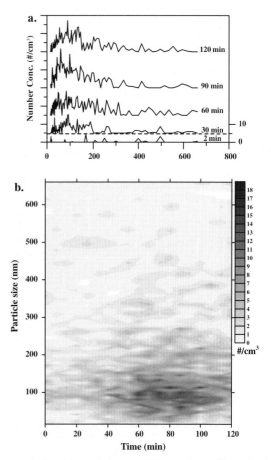

Figure 2. Particle emission at different times (a) and 2-h scale (b) of TiO_2/wood using UV light, fan and scraping motion (case 1).

size distribution of the particles released from TiO₂/wood for the whole 2-h test. Most of the particles are between 50 and 150 nm. The highest number concentration is about 18 $\#/cm^3$ at the particle size of 120 nm. After 40 min of operation, a noticeable increase in the particle concentration was detected. The results indicate that the conditions in case 1 can increase the number concentration of particle emission from TiO₂ nanopowder coating material. For case 2 (without scraping motion), the particle number concentration is less than that in case 1 but the size distribution is similar. In the simulated experiments, the scraping motion (movement of the rubber knife) is believed to remove TiO₂ nanopowder from the surface of coating materials as the UV light

deteriorates organic compound used in the coating agent. Although the simulation is a worse scenario compared to real applications, nanoparticle emission from utilization of TiO₂ nanomaterial should not be neglected.

Effect of light source

UV light is the most important parameter for photo catalyst products applied on the outside of buildings. UV and visible light were simulated in this study using UV and fluorescent lamps (case 1 and 3). Similar to Figure 2, Figure 3a, b show the measurement result of TiO₂/wood in the simulation case 3. The particle number concentration is below 3 $\#/cm^3$ in this case and hardly seen in the figure. The results suggest that the scraping motion of the rubber knife does not cause too much particle emission under the fluorescent lamp light. Comparing Figure 2 with Figure 3 for particle emission, the differences indicate that the destructive effect of UV light on the coating agent is much more serious than the one of fluorescent lamps. UV light with high energy is capable of reducing the binding force between TiO₂ nanopowder and the substrate surface, therefore, many particles are released as shown in Figure 2.

Substrate effect

Figure 4 shows the particle emission of TiO₂/polymer in the simulation case 1. After 30 min of operation, a large amount of particles below 200 nm were released and detected. The highest number concentration of particle emission was about 16 $\#/cm^3$ at the particle size of 100 nm. Comparing Figure 4 with Figure 2, the results show that the number of particles being emitted increased as the testing time reached a critical point. The breaking time of significant particle emission is about 30 min for TiO₂/polymer and 40 min for TiO₂/wood. After the simulated experiment, the color of the polymer changed from white to yellow. The color change indicates that the stability of TiO₂ nanopowder coating on the polymer is worse than on wood. In the simulating process, UV light may have caused a property change in the polymer substrate and/or activated the TiO₂ photo catalyst to change the properties of the binder and/or the polymer substrate.

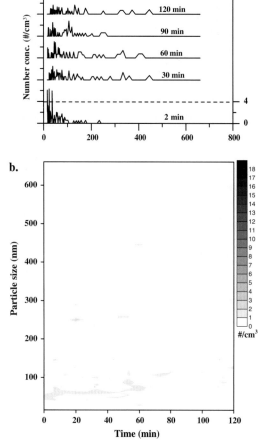

Figure 3. Particle emission at different times (a) and 2-h scale (b) of TiO₂/wood using fluorescent lamp, fan and scraping motion (case 3).

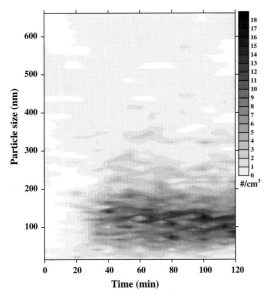

Figure 4. Particle emission of TiO$_2$/polymer using UV lamps, fan and scraping motion (case 1).

Figure 5 shows the measurement results of TiO$_2$/tile. After 30 min of operation, a large amount of particles below 200 nm were released and detected. The highest number concentration of particle emission is about 22 #/cm^3 at the particle size of 55 nm. The increase of particle emission after the critical time, is similar to the results in Figure 4. In addition to the UV light and rubber knife scraping motion, the binder used in the commercial TiO$_2$ paint, surface properties of the substrate and coating skills are also important to influence the life cycle and properties of photo catalyst products.

Table 2 summarizes the evaluation results of particle emission showing the test conditions, running time, number median diameter (NMD), number concentration and average emission rate. The results show that, under the conditions of UV lamps, a fan and scraping motion, the particle number concentration or average emission rate decreases significantly after 60 and 90 min for TiO$_2$/polymer and TiO$_2$/wood, respectively. However, the emission rate is still increasing after 2 h of testing for TiO$_2$/tile. According to the test results, evaluation by simulation experiments is useful for assessing possible particle emission for nanomaterial coating products and to reduce the exposure likelihood of customers who are using or applying the nanopowder coating products.

Conclusions

This study has investigated the possibility of particle emission from TiO$_2$ nanopowder coating materials under several simulated conditions including UV light/fluorescent lamps, a fan and rubber knife scraping motion. Among the three selected substrates (wood, polymer and tile), tile coated with TiO$_2$ nanopowder was found to have the highest particle emission (22 #/cm^3 at 55 nm). UV light was shown to increase the release of particles below 200 nm from TiO$_2$ coating products. Another important factor was the rubber knife scraping motion. These actions together can greatly reduce the binding force between the TiO$_2$ nanopowder and the substrate surface and, therefore, produce particle emissions from the coating products. The particles, released from nano-products to the ambient environment, may potentially increase the exposure possibility of consumers. It is suggested that particle emission evaluation should be carried out for nanopowder coating materials and studies of surface preparation, functional

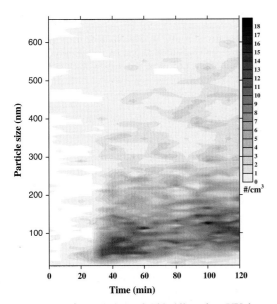

Figure 5. Particle emission of TiO$_2$/tile using UV lamps, fan and scraping motion (case 1).

Table 2. Summary of particle emissionat different operating conditions

Test conditions	Time (min)	Number median diameter (NMD) (nm)	Number conc. (#/cm^3)	Average emission rate (#/min)
TiO$_2$/Wood using UV light, f an and scraping motion	2	213.7	25.3	11.2
	30	53.5	231.5	102.9
	60	8.1	458.3	203.7
	90	20.9	633.3	281.5
	120	38.4	478.5	212.7
TiO$_2$/Wood using fluorescent lamp, fan and scraping motion	2	35.5	64.7	28.8
	30	61.3	40.1	17.8
	60	63.8	40.5	18.0
	90	83.6	31.0	13.8
	120	74.7	24.6	10.9
TiO$_2$/polymer using UV lamps, fan and scraping motion	2	93.1	29.8	13.2
	30	94.4	460.3	45.4
	60	97.5	449.8	199.9
	90	102.1	383.7	170.5
	120	102.7	378.9	168.4
TiO$_2$/tile using UV lamps, fan and scraping motion	2	36.5	12.4	5.5
	30	107.3	448.8	199.5
	60	127.1	442.2	196.5
	90	127.7	525.3	233.5
	120	119.6	629.2	279.6

clothes, and nano-composite filters should also be performed.

References

Aitken R.J., K.S. Creely & C.L. Tran, 2004. Nanoparticles: an occupational hygiene review. Institute of Occupational Medicine. Res. Rep. 274, 12–15.

Chen T.M. & H.M. Chein, 2003. Nanoparticle generation and monitoring. Center for Environmental Safety & Health Technology Development, Industrial Technology Research Institute, pp. 58–78 (in Taiwan).

Elder A.C.P., R. Gelein, M. Azadniv, M. Frampton, J. Finkelstein & G. Oberdorster, 2002. Systemic interactions between inhaled ultra fine particles and endotoxin. Ann. Occup. Hyg. 46(Suppl. 1), S231–S234.

Elder A.C.P., R. Gelein, M. Azadniv, M. Frampton, J. Finkelstein & G. Oberdorster, 2004. Systemic effects of inhaled ultra fine particles in two compromised, aged rat strains. Inhal. Toxicol. 16(6/7), 461–471.

Elder A.C.P., R. Gelein, J.N. Finkelstein, C. Cox & G. Oberdörster, 2000. Pulmonary inflammatory response to inhaled ultra fine particles is modified by age, ozone exposure, and bacterial toxin. Inhal. Toxicol. 12(Suppl. 4), S227–S246.

Ferin J., G. Oberdörster, S.C. Soderholm & R. Gelein, 1991. Pulmonary tissue access of ultra fine particles. J. Aerosol Med. 4(1), 57–68.

Ferin J. & G. Oberdörster, 1992. Translocation of particles from pulmonary alveoli into the interstitium. J. Aerosol Med. 5(3), 179–187.

Kreyling W., M. Semmler, F. Erbe, P. Mayer, S. Takenaka, H. Schulz, et al., 2002. Translocation of ultra fine insoluble iridium particles from lung epithelium to extra pulmonary organs is size dependent but very low. J. Toxicol. Environ. Health 65A(20), 1513–1530.

Li X., D. Brown, S. Smith, W. MacNee & K. Donaldson, 1999. Short-term inflammatory responses following intratracheal instillation of fine and ultra fine carbon black in rats. Inhal. Toxicol. 11, 709–731.

Nemmar A., A. Delaunois, B. Nemery, C. Dessy-Doize, J.F. Beckers, J. Sulon, et al., 1999. Inflammatory effect of intratracheal instillation of ultra fine particles in the rabbit: role of C-fiber and mast cells. Toxicol. Appl. Pharmacol. 160, 250–261.

Nemmar A., P.H.M. Hoet, B. Vanquickenborne, D. Dinsdale, M. Thomeer, M.F. Hoylaerts, et al., 2002a. Passage of inhaled particles into the blood circulation in humans. Circulation 105, 411–414.

Nemmar A., M.F. Hoylaerts, P.H.M. Hoet, D. Dinsdale, T. Smith, H. Xu, et al., 2002b. Ultra fine particles affect experimental thrombosis in an in vivo hamster model. Am. J. Respir. Crit. Care Med. 166, 998–1004.

Nemmar A., M.F. Hoylaerts, P.H.M. Hoet, J. Vermylen & B. Nemery, 2003. Size effect of intratracheally instilled particles on pulmonary inflammation and vascular thrombosis. Toxicol. Appl. Pharmacol. 186, 38–45.

Oberdörster G., J. Ferin, R. Gelein, S.C. Soderholm & J. Finkelstein, 1992a. Role of the alveolar macrophage in lung injury: studies with ultra fine particles. Environ. Health Persp. 97, 193–197.

Oberdörster G., J.N. Finkelstein, C. Johnston, R. Gelein, C. Cox, R. Baggs et al., 2000. HEI Research Report: Acute Pulmonary Effects of Ultra fine Particles in Rats and Mice HEI Research Report. No. 96: Health Effects Institute.

Oberdörster G., R.M. Gelein, J. Ferin & B. Weiss, 1995. Association of particulate air pollution and acute morality: involvement of ultra fine particles? Inhal. Toxicol. 7, 111–124.

Oberdörster G., Z. Sharp, V. Atudorei, A. Elder, R. Gelein, W. Kreyling, et al., 2004. Translocation of inhaled ultra fine particles to the brain. Inhal. Toxicol. 16(6/7), 437–445.

Oberdörster G., Z. Sharp, V. Atudorei, A. Elder, R. Gelein, A. Lunts, et al., 2002. Extra pulmonary translocation of ultra fine carbon particles following whole-body inhalation exposure of rats. J. Toxicol. Environ. Health 65A, 1531–1543.

Oberdörster G., A. Maynard, K. Donaldson, et al., 2005. Principles for characterizing the potential human health effects from exposure to nanomaterials: elements of a screening strategy. Part. Fibre Toxicol. 2, 8.

Roco M.C., 2005. International perspective on government nanotechnology funding in 2005. J. Nanopart. Res. 7, 707–712.

Zhou Y.-M., C.-Y. Zhong, I.M. Kennedy, V.J. Leppert & K.E. Pinkerton, 2003. Oxidative stress and NFkB activation in the lungs of rats: a synergistic interaction between soot and iron particles. Toxicol. Appl. Pharmacol. 190, 157–169.

Journal of Nanoparticle Research (2007) 9:165–182
DOI 10.1007/s11051-006-9151-0

Special focus: Nanoparticles and Occupational Health

Moving forward responsibly: Oversight for the nanotechnology-biology interface

Jennifer Kuzma[1,2]
[1]*Center for Science, Technology, and Public Policy, University of Minnesota, Minneapolis, MN, USA;* [2]*254
Humphrey Center, 301 19th Avenue South, 55455, Minneapolis, MN, USA; *Author for correspondence
(Tel.: +612-625-6337; Fax: +612-625-3513; E-mail: kuzma007@umn.edu)*

Received 24 July 2006; accepted in revised form 28 July 2006

Key words: oversight, regulatory policy, nanotechnology, biotechnology, biology, societal implications

Abstract

Challenges and opportunities for appropriate oversight of nanotechnology applied to or derived from
biological systems (nano-bio interface) were discussed in a public workshop and dialog hosted by the
Center for Science, Technology, and Public Policy of the University of Minnesota on September 15, 2005.
This paper discusses the themes that emerged from the workshop, including the importance of analyzing
potential gaps in current regulatory systems; deciding upon the general approach taken toward regulation;
employing non-regulatory mechanisms for governance; making risk and other studies transparent and
available to the public; bolstering mechanisms for public participation in risk analysis; creating more
opportunities for meaningful discussion of the social and ethical dimensions of the nano-bio interface;
increasing funds for implications and problem-solving research in this area; and having independent and
reliable sources for communication. The workshop was successful in identifying ways of moving forward
responsibly so that ultimately nanotechnology and its products can succeed in developers', researchers',
regulators', and the public's eyes.

Introduction

Nanotechnology, or really any technology, does
not exist in a vacuum. It is derived from human
efforts and affected by social, cultural, and political climates. Yet, few technologies emerge with the
societal consequences in mind. Social and economic issues often gain most attention after technologies enter the marketplace and are widely
used. However, within the U.S. National Nanotechnology Initiative (NNI), resources have been
directed toward the investigation of the societal
issues that might accompany the applications of
nanotechnology on their road to development and
use. Approximately 4% of NNI funding has been
used to study the social, educational, and ethical

implications of nanotechnology (Roco, 2005).
Many experts in this area have cited negative
experiences with past technologies, such as stem
cell research and genetic engineering, as the
impetus for dealing with the contextual issues for
nanotechnology early and often.

Society drives and regulates technology,
attempting to minimize the downsides and maximize the benefits. Appropriate oversight of new
technologies is important for ensuring the health
and environmental safety of products and instill-
ing public confidence. Most people agree that
ultimately the success of any technology is
dependent on proper oversight within a societal
context. Mishaps or accidents can preclude future
use and development, and there is a delicate

balance between allowing technology to flourish and putting appropriate oversight mechanisms in place.

Discussions of oversight frameworks for nanotechnology have largely focused on health issues associated with engineered nanoparticles or those in cosmetics, such as carbon nanotubes, buckyballs, and titanium dioxide (e.g. Long et al., 2006).[1] Less attention has been paid to oversight for widespread applications in medicine, food and agriculture, and the environment, for which patients, ecosystems, farmers, or the general public may bear the risks and benefits. Several of these "nano-bio" applications are already entering the marketplace, while others are emerging into development and clinical trial phases. However, there have not been many focused public conversations on appropriate oversight frameworks for them. It is in this context, that the Center for Science, Technology, and Public Policy hosted the workshop "The Nanotechnology-Biology Interface: Exploring Models for Oversight" on September 15, 2005.[2] The workshop addressed the following questions:

- Is there or will there be a mis-match between the ability to generate nanoparticles and the ability to detect or determine the effects of these particles? Should the two be linked in any regulations developed with respect to nanoparticle use?
- Are there procedures developed for other technologies that could or should be adopted or adapted to assure the safety of nanoparticles and materials developed from them? What is the appropriate balance between government regulation and investigator or industry voluntary guidelines?
- What is the relationship between claims (from modest to extravagant) made for the potential of

nanotechnology and the challenge of building public confidence in the safety of the new technology? What is the appropriate strategy for balancing these two factors to preserve momentum in the development of the technology?
- If new governance models are designed, or existing ones revised, what is the appropriate process? What scientific, economic, social, and other factors should be considered? Who should be involved in developing the models?

This paper highlights key points of discussion from the workshop and the opportunities and challenges associated with oversight at the nano-bio interface.[3] It represents the opinions of the delegates at the workshop[4] and additional information and conceptual synthesis by the author, but it does not represent the consensus of the participants nor the views of the organizations with which they are affiliated. The paper is organized around themes of the workshop sessions: science and applications at the nano-bio interface, health and environmental safety concerns, regulatory and non-regulatory approaches to governance, striking an appropriate balance for oversight in a societal context, and far-future applications and how governance systems can account for them. It concludes with a summary of the conclusions and recommendations that emerged from the dialog. The author of this paper believes that it should be just the beginning of closer examinations of oversight for the nano-bio interface.

Science and applications

Some question how nanotechnology came to be and whether it is really new. Material sciences, biochemistry, chemical engineering, aerosol science, and particle technology have been around for several decades, yet have only recently been included in the forefront of nanotechnology research and development. To better understand the field of nanotechnology, the first session of the workshop was designed to review specific applications of nanotechnology to biology (Table 1)

---

[1] 2nd International Symposium on Nanotechnology and Occupational Health, University of Minnesota, October 3–6, 2005.

[2] In a public workshop, over 160 people from academe, industry, state and federal government, trade organizations, law and venture capital firms, and the general public convened to discuss governance issues for the nano-bio interface. The day following the public workshop, approximately 35 speakers and other experts continued the discussions. The workshop was funded by the Consortium on Law and Values in Health, Environment, and the Life Sciences. http://www.lifesci.consortium.umn.edu/

[3] For a more complete summary of the presentations, visit http://www.hhh.umn.edu/centers/stpp/nanotechnology.html.

[4] Citations in the text with an asterisk represent speaker presentations and not peer reviewed publications.

Table 1. A few examples of research and applications at the nano-bio interface

Sector	Application	Method or material	Details
Agriculture	Basic research on energy production	Nanodetection	Single molecule detection to determine enzyme/substrate interactions (e.g. cellulases in production of ethanol).
	Agrochemical delivery	Nanoparticles, nano-capsules	Delivery of pesticides, fertilizers, and other agrichemicals more efficiently (e.g. only when needed or for better absorption).
	Animal production	Nanoparticles	Delivery of growth hormone in a controlled fashion.
		Nanomaterials in chips (nanochips)	Identity preservation and tracking.
	Animal or plant health	Nanosensors	Detect animal pathogens, such as foot and mouth disease virus. Detect plant pathogens early.
	Animal medicine	Nanoparticles, nanodevices	Deliver animal vaccines.
	Plant production	Nanoparticles	Delivery of DNA to plants toward certain tissues (i.e. targeted genetic engineering).
Food	Sensing	Nanosensors	Detect chemicals or foodborne pathogens; biodegradable sensors for temperature, moisture history, etc.
	Safety	Nanoparticles	Selectively bind and remove chemicals or pathogens.
	Packaging	Nanoclays, nanofilms	Prevent or respond to spoilage. Sensing features for contaminants or pathogens.
	Healthy food	Nanoemulsions, nanoparticles	Better availability and dispersion of nutrients, nutraceuticals, or additives.
Environment	Microbial ecology and characterization	Microfluidics, micro-nano arrays	Identify and quantify microbial populations for biocontrol, composting, bioremediation, etc. (e.g. nano-single strand conformation profiling of DNA for counting single molecules).
	Sensing	Nanosensors	Detect environmental contaminants early; assess states of populations or ecosystems.
	Remediation	Nanoparticles	Bind contaminants and remove them.
Medicine	Ex vivo diagnostics	Nanoparticle labeled DNA	Microarray analysis for medical genomics (e.g. detecting patient genetic response to pathogens, or tailoring treatments to individuals).
	In vivo diagnostics	Nanoparticles	Magnetic particles for imaging (e.g. MRI of tumors).
	Drug delivery or gene therapy	Nanoparticles	Tag particles and target drugs or genes to specific tissues. Increase bioavailability or solubility. Make drugs more bioactive. Liposomes, dendrimers, and other inorganic or organic particles.
	Tissue scaffold	Surface nanomaterials, nanocoatings	Promote cell growth, providing a matrix (e.g. neuron growth on nanofabricated silicon).
	Medical devices	Nanocoatings	Coatings to make devices, such as pacemakers, corrosion resistant and biocompatible.
		Nanosensors	Sense chemicals in the body and adjust device function accordingly.
Biology – Basic research	Sensing, detecting, and characterizing	Cantilevers	Bend in response to molecular interactions.
		Nanowires	Use nano-conductivity for characterizing interactions (e.g. between viruses and proteins).
		Quantum dots	Label different cells, proteins, or cellular components with colored tags. Track movement, relationships, etc. at the subcellular level. Long lifetime is a positive feature of these tags.

Note: this table does not include all possible categories of applications. *Source*: compiled by J. Kuzma.

and tackle the question of what makes nanotechnology special.

Skeptics believe that as a society we have hijacked many different fields and applications and now call it "nanotechnology" as a way to "create buzz" and get funding. They argue that nanotechnology is not entirely new, and therefore, in public policy discussions, it is important that we do not forget that the basic scientific principles associated with nanotechnology are the same as with other technologies. However, others would argue that in last 10 years, we have had better ability to control, understand, and analyze materials at the nanoscale and that nanomaterials have special properties. Therefore, a new area is justified.

There is disagreement as to whether a precise definition of nanotechnology is necessary for oversight and framing of other contextual issues. Nanotechnology is defined by the NNI as "the understanding and control of matter at dimensions of roughly 1–100 nanometers, where unique phenomena enable novel applications" (NNI, 2005). Nanomaterials can arise from a "top-down" approach, in which macroscopic material is broken down to the nanoscale, or from a "bottom-up" approach, in which individual atoms or molecules are coaxed or self-assemble into nanoparticles. Regardless, nanotechnology reflects a multidisciplinary conglomerate of ideas and methods. In this sense, it is an umbrella that brings people together to make discoveries and solve problems. Secondary effects of bundling various basic research questions, fields, disciplines, and applications together include increased collaboration and understanding among the actors and scientists involved.

Nanomolecules and particles, such as viruses, molecular motors, and membrane vesicles, exist naturally in biology, but the new nano-bio interface explores the interactions of the natural with the man-made (Taton, 2005*). The nano-bio interface is here, and the pace of applying nanotechnology to health, environment, agriculture, and food is moving quickly. This rapid emergence creates a host of technical challenges and opportunities, as the nanoscale is unique from a physical standpoint. For example, workshop participants discussed whether quantum dots interact with molecules in cells and interfere with typical cellular reactions. There are studies which indicate that quantum dots do bind cellular components, and researchers are working to develop biocompatible coatings to prevent such interactions (Taton, 2005*).

For industry, a primary question is what can be done with nanotechnology to benefit society. The medical device industry uses nanotechnology to make corrosion-resistant coatings. In the future, it plans to develop nanosensors that can be used to detect physiological states of patients (Untereker, 2005*). Nanotechnology can also be used in agriculture and the environment, for example, to better understand how cellulases work to produce ethanol and identify and quantify naturally-occurring microorganisms for generating products and energy from waste (Table 1) (Walker, 2005*).

Health and environmental safety

The ability to assess the health and environmental impacts of nanoparticles and other nanomaterials is a cornerstone of governance. It is not sufficient, but seems necessary for a good oversight framework. Standard models of chemical, ecological, or microbial risk assessment apply to nano-bio products, but within the models, data needs and technical questions might be very different given the special biological, chemical, and physical properties that arise at the nanoscale.

At the workshop, current activities and studies in environmental health and safety (EHS) research were presented, along with appropriate mechanisms and institutions for funding and conducting such work. Several federal agencies fund EHS research under NNI, such as the National Science Foundation (NSF), Environmental Protection Agency (EPA), National Institutes of Health (NIH), Department of Energy (DOE), Department of Defense (DOD), the National Institute of Occupational Safety and Health (NIOSH), and National Institute of Standards and Technology (NIST). For fiscal year (FY) 2006, the NNI EHS research budget request totaled $38.5 million out of the total NNI budget request of $1 billion (NSET, 2005; Savage, 2005*). Yet, there is general difficulty in classifying EHS research as "implications" or "applications," and this is a point of contention in the health and environmental safety community. For example, the general development of sensors could be placed in either category. Also, applications research can provide information for implications and vice versa. To date, EPA has funded approximately $15.6 million of applications research, to address existing environmental problems or prevent future problems, and it has funded $10.2 million in implications research, to address the interactions of nanomaterials with the environment and potential risks (Savage, 2005*). Environmental applications include improved monitoring and detection capabilities, ultra-green manufacturing and chemical processing, waste minimization, reduced energy usage, clean energy sources, remediation and treatment technologies, and sustainability applications. Implications research for nanomaterials includes studies on toxicity and its mechanisms; effects of manufacturing on ecosystems; transportation and fate of nanomaterials; bioaccumulation, transformation, and availability; and dose–response assessment.

Many health and environmental applications of nanotechnology have a dual nature. The novel properties and uses of nanotechnology which provide benefits also cause regulatory concern. For example, the ability of nanoparticles to circumvent the blood–brain barrier (Oberdörster et al., 2005a) has advantages for delivering drugs to the brain, which is an organ that is otherwise difficult to reach, but that same feature amplifies neurotoxicity concerns. In environmental applications, nanomaterials may be used to penetrate and remediate subsurface areas, but their penetration abilities could also lead to greater ecosystem damage. Surface properties are much different at the nanoscale, and quantum properties dominate. Increased surface areas of nanoparticles can lead to increased cellular reactivity, and increased bioavailability to increased toxicity. Lower effective doses of nanomaterials could mean toxic effects at lower doses. Greater abilities of nanoparticles to penetrate membranes and speed the onset of action could lead to greater toxic exposure through non-traditional routes (Finkelstein, 2005*; Hoerr, 2005*). At the nanoscale, cellular uptake mechanisms are different – particles below 20 nm can be taken up by the endothelium skin layers and those below 10–50 nm can enter cells through receptor mechanisms (Finkelstein, 2005*). The special properties and effects of nanoparticles need to be considered, and new strategies for toxicology are needed.

Fundamental knowledge and data for assessing the potential risks of nano-bio products is lacking.

Some factors complicating risk assessments include differences in the exposure medium (e.g. air, water, or food), routes of exposure (e.g. inhaled, consumed, or contacted), and dose–response relationships. In some cases, it is not known whether nanomaterials exist as single particles or are agglomerated in certain media. Also, there are differences in dose–response curves depending on whether the curves are expressed by mass, number of particles, or surface area (Finkelstein, 2005*). Furthermore, most tests are short-term, and long-term toxicity and effects remain unknown. Key questions for toxicology research on nanomaterials include: which physicochemical characteristics of nanoparticles are associated with adverse effects (e.g. size, chemistry, crystallinity, biopersistence, surface coating, porosity, charge)?; is cellular uptake involved, and if so, what are the uptake and translocation mechanisms?; and what should be considered when designing biocompatible nanomaterials? (Finkelstein, 2005*). An expert group of the International Life Sciences Institute Research Foundation recently published a report on a screening strategy for hazard identification of nanomaterials based on their characteristics (Oberdörster et al., 2005b).

Until more data are generated, understanding from previous fields could be applied nanotechnology. Nanotechnology is a broad field, for example, covering particles in films, surfaces, liquids, and gases, and developing methodology for study of the environmental and health effects is hard to do as a result. It could be beneficial to select focused systems that are well-established, such as aerosol science, in order to better understand nanoscience and its applications (Friedlander, 2005*). Aerosol technology has a large domain, spanning particle diameters from 10^{-3} to 10^{2} μm (0.1 to 100,000 nm), and the industry has had many years of experience in data collection, including for worker exposure, and particle characterization. The aerosol system provides a useful example for developing powerful methodology for nanotechnology (Friedlander, 2005*).

Governance frameworks

Governance systems are currently in place for a variety of products associated with new technologies. For the design of oversight models for nanotechnology, there are opportunities to learn from past experience with other technologies and products. It is important to not recreate what already exists. Some argue that nanotechnology is already suitably covered by existing regulatory and non-regulatory oversight activities, whereas others disagree, arguing that many products on the market are falling through the cracks of a system that has not been formalized or coordinated. As a society, we have choices of (1) creating new laws and/or regulations, (2) revising existing ones, (3) interpreting existing ones to cover nanoproducts, (4) designing non-regulatory approaches, or (5) modifying existing non-regulatory approaches. The diversity of the products at the nano-bio interface (Table 1 and 2) might preclude a single approach or framework, and one size might not fit all. The second session of the workshop examined existing or possible approaches for nanotechnology applied to or derived from biological systems. It also considered appropriate features of governance systems for ensuring public confidence and safety.

One example of oversight from which lessons could be learned is the Coordinated Framework for the Regulation of Biotechnology (OSTP, 1986). In this case, a governance approach was developed for the products of biotechnology by using a patchwork of existing laws. EPA, the Food and Drug Administration (FDA), and the United States Department of Agriculture (USDA) were identified as lead agencies for specific products. Underlying principles of the framework were that the products, not the process of biotechnology, should be the focus of regulation and that genetically engineered organisms (GEOs) are not fundamentally different from non-engineered organism. Therefore, existing laws were determined to be sufficient. After the framework was published, the agencies chose paths to develop regulations under existing laws or provide guidance and policies under them. This approach and others should be closely examined for their relevance to the products of nanotechnology.

EPA and its role in nano-bio oversight were examined at the workshop. The budget of EPA has been flat or decreasing in recent years, and diminished resources pose difficult challenges for EPA in prioritizing its activities (Bergeson, 2005*). Despite the challenges, the agency is considering how to reap the benefits of nanotechnology and identify and control the risks. EPA convened its Science Advisory Board (SAB) in December 2004 to discuss nanotechnology. The SAB concluded that nanotechnology can provide great benefits, but nanomaterials require additional investigations of environmental, health, and social impacts. SAB members noted that advancements in new technologies are occurring at unprecedented rates, making it difficult for government agencies to keep abreast of emerging developments.

EPA is also providing general guidance on the scope of its regulatory authority under the Toxic Substances Control Act (TSCA) and considering a voluntary program for the review of existing nanomaterials. Amending TSCA is unlikely in the short term, and perhaps even in the long term, therefore, creative ways are needed to ensure that nanoscale materials are addressed appropriately under existing legal authorities (Bergeson, 2005*). The National Pollution Prevention and Toxics Advisory Committee (NPPTAC) was formed under the auspices of the Federal Advisory Committee Act to provide guidance to EPA on TSCA implementation and related EPA toxics programs. The NPPTAC agreed to form an *ad hoc* interim work group on nanoscale materials to provide guidance on the prudence and scope of a voluntary reporting program on nanoscale materials. The Nanoscale Materials Voluntary Program (NVP) was agreed upon in concept during NPPTAC *ad hoc* interim work group discussions in summer 2005, which progressed and concluded in fall 2005.

Details of the NVP for which agreement will be difficult include defining its scope (e.g. should it review emerging chemicals, or only those now in commerce), deciding whether data generation should be part of the program, balancing transparency with protecting confidential business information, and including a diverse number of small and medium sized enterprises. The NPPTAC transmitted to EPA Administrator Johnson on November 22, 2005, its "Overview of Issues for Consideration by NPPTAC" (EPA, 2005a). The document offers the NPPTAC's "analysis and views of a framework for EPA's approach to a voluntary program for engineered nanoscale materials, a complementary approach to new chemicals nanoscale requirements under the Toxic Substances Control Act (TSCA), and other relevant issues presented."

Table 2. A few examples of products of nanotechnology applied to health, food, or the environment

Company	Product	Purpose/method	Stage
Health/medicine			
American Bioscience, Inc. (ABI)	Abraxane™	Nanoparticulate formulation of the widely used anticancer drug paclitaxel for metastatic breast cancer. First approval of protein albumin nanoparticles as a "natural solvent."	Received final FDA approval in Jan. 2005.
BioSante pharmaceuticals	BioVant™, BioOral™, BioAir™	Calcium phosphate-based (CAP) nanotechnology for oral, nasal, and trans cutaneous routes of delivery of vaccines and proteins.	CAP is in preclinical safety trials, as indicated by the company website.
GP Surgical	TiMESH	Hernia mesh made with titanium nano-coating.	FDA approved, implanted device; commercially available.
Health Plus International, Inc.	Spray For Life®, Vitamin B12 Energy Booster	Nanoceutical™ Delivery System (NDS), disperses active molecules into nanodroplets, increases bioavailability of nutrients or drugs.	Commerically available, but not FDA approved.
StarPharma	VivaGel™	Polyvalent, polylysine dendrimer as the active ingredient. Intended to prevent transmission of sexually transmitted diseases (STDs).	Determined to be safe and well-tolerated in Phase 1 clinical trials as an Investigational New Drug (IND) under FDA (2004), according to the company website.
Environment/health			
EnviroSystems	EcoTru®	Nanoemulsion technology to disinfect surfaces for bacteria and viruses.	Commercially available.
Severn Trent Services & Bayer AG	SORB 33® Bayoxide® E33	Nano-sized surface structures that are able to absorb arsenic; composed of ferric oxide.	Certified for drinking water systems by American National Standards Institute/ NSF Standard 61. FDA pre-market review would be required under the Federal Food Drug and Cosmetic Act (FFDCA) if used in bottled water.
Food/health			
bioMerieux	FoodExpert-ID®	High-throughput gene chip for testing food and animal feed for traceability and safety.	In trials in some European countries (2004).
Nanocor	Variety of products under the Nanomer® trademark	Nanoclays and composites providing barriers to oxygen and carbon dioxide flow used in food packaging to keep freshness and block out smells.	Nanomer® nanoclays are available for commercial use. Infrastructure in place to produce more than 100 million pounds annually.

Table 2. Continued

Company	Product	Purpose/method	Stage
Nanoplex Technologies	Nanobarcodes® Particles	Encodeable, machine-readable, durable, metallic rods. Particles are intrinsically encoded by virtue of the difference in reflectivity of adjacent metal stripes. Used for supply-chain tracking for food.	Product expected on market in 2006.
NutraLease Shemen Industries Ltd	Canola Active	Nanocapsules in cooking oil to improve bioavailability of nutraceuticals, for example, plant sterols to reduce the body's absorption of cholesterol in the blood.	On market in Israel; FDA has not reviewed this product. Sponsor website indicates "Canola Active – complies with FDA requirements."
OilFresh	OilFresh™	Vertical insert, made of an advanced nanoceramic material. For use in cooking oil for better quality.	Sponsor website indicates "OilFresh is authorized by the FDA." However, as material is not expected to migrate into food, FDA pre-market review was not required.
Samsung	Nano SilverSeal™	Nano-silver compound in the product design to suppress the spread of bacteria and other microbes in refrigerators.	Commercially available.

Note: this is not a complete list. *Source*: J. Kuzma, compiled from company web searches, and many examples from the list were provided by N. Savage, EPA.

In addition to the above activities, EPA's Science Policy Council (SPC) recently released a white paper on nanotechnology that will have cross-program implications for the agency (EPA, 2005b). The paper reviews science deficits and data needs for the agency and guides programmatic elements for both applications and implications of nanotechnology Overall, challenges for EPA are to stay on top of program priorities, set years in advance, while responding to crises and anticipating and managing emerging technologies, including nanotechnology (Bergeson, 2005*). Training, recruitment, and infrastructure development were identified by the SAB as key priorities for EPA to meet these challenges.

FDA's approach to nanotechnology is similar to its approach for any other technology, as the agency regulates on a product-by-product basis. It does not regulate technologies *per se*, but rather several kinds of products in various ways (Alderson, 2005*). Drugs are regulated via a pre-market approval process – they have to pass safety, efficacy, and manufacturing standards. Devices that are of low risk are regulated on the basis of "acceptance of the product." They are expected to meet pre-approved standards (category 510ks), and if these are met, marketing can begin. Cosmetic products can be marketed without FDA evaluation or review.

Safety considerations for FDA include access of nanomaterials to cells and tissues, time and clearance of materials in cells and tissues, and effects on cell and tissue function. Absorption, distribution, metabolism and excretion (ADME) of nanomaterials are key issues. Extensive pre-clinical tests, such as pharmacology, toxicology, geno-toxicity, developmental toxicity, immunotoxicity, and carcinogenicity tests, are required for safety evaluations. Features of tests include the use of high dose multiples, at least two animal species, histopathology on most organs, and extended dosing periods. However, FDA does not conduct these safety tests themselves, and the drug sponsor develops them. The agency provides guidance and direction for the tests and evaluates them after they are conducted. It considers not only safety, but also pre-clinical medical utility for products and manufacturing standards for consistent and quality products. In order to evaluate products, the agency needs more information on the forms of nanoparticles that are presented to cells, the stability and critical properties of nanomaterials, and the effects of scale-up and manufacturing on the characteristics of nanomaterials (Alderson, 2005*).

There are limits to FDA authority and regulatory activity. FDA can only regulate products based on the claims of the sponsor, and ultimately, FDA may be unaware that nanotechnology is being used in a particular product. Furthermore, FDA has only limited authority for some nanoproducts, such as cosmetics. Four key challenges for FDA were presented at the workshop: the need to (1) be prepared for "unknown" risks, deal with them, and adopt new procedures for doing so, (2) communicate with manufacturers of new medical products, (3) involve stakeholders, (4) communicate risks to the public,

and (5) report relevant scientific findings in a timely fashion (Alderson, 2005*).

The public and industry are concerned about controlling the technology, managing the risks, and considering potential gaps in regulation (Michelson, 2005*). Workers in industry and students in academic labs are the primary ones who are exposed to nanoparticles right now, and so are those in developing countries where infrastructure and training for health and environmental safety are lacking (Michelson, 2005*). Both in the U.S. and abroad, there is no agreed upon guidance in terms of worker safety practices for nanomaterials. Regulatory guidance is particularly important for small businesses that often do not have the resources to devote to environmental health and safety.

Risks will change and shift focus as the technology does. We are currently moving from the development and use of passive nanostructures, such as coatings and polymers, to active nanostructures, such as targeted drugs and other

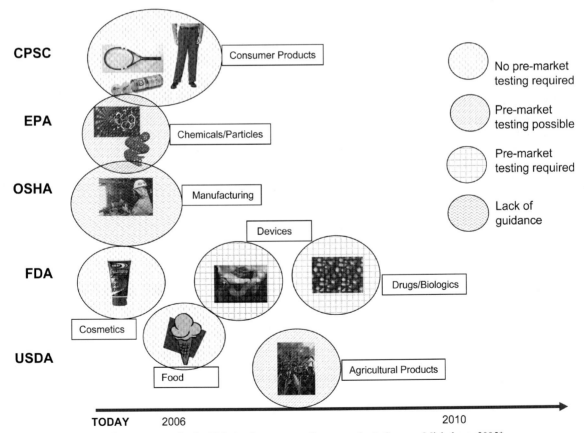

Figure 1. Current regulatory system in the U.S. in the context of nanoproducts Source: Michelson, 2005*.

applications at the nano-bio interface (Roco, 2004). As the technology moves forward, people will only accept risks if important benefits occur. However, many are worried that the benefits, such as cancer treatment or cheap and clean energy, will not materialize if regulations are not in-step with advances and a mishap occurs as a result. With a growing number of nano-based products out on the market, the federal oversight process will increasingly have trouble keeping up with the pace of product development and market entry, as it can take several years to fund and conduct research on the health and environmental risks, and even longer to amend or formulate regulations (Michelson, 2005*).

It is important to consider current and potential cracks in the regulatory system for nano-bio products, as well as the general approach that we, as a society, want to take toward regulation. Some products make it through the system today without required safety reviews (Figure 1 and Table 2). For some substances, like drugs, there is comprehensive federal safety review, and other products fall in between these two extremes. There are a range of potential approaches to the regulation of nanomaterials, from more restrictive, like a moratorium on use or the treatment of nanomaterials as new substances, to more permissive, like voluntary standards and guidelines or no special state of regulation. At the workshop, an example of proactive regulation was discussed. In 1976, Cambridge MA passed a city ordinance for using and reviewing recombinant DNA research, and this was one of the factors that led the biotechnology company Biogen, Inc. to move its headquarters there (Michelson, 2005*). Public aspects of transparency, good governance, and a mature understanding of the field can make locations attractive for businesses.

Non-regulatory schemes for governance could also be important for the nano-bio interface. Tort law and insurance were discussed at the workshop. The primary function of tort law is to remedy harm or compensate victims, while that of insurance is to spread risk (Ziegler, 2005*). In some cases, a defect in manufacturing, design, or warning is necessary for product liability. However, in other cases, an injury traceable to the product used in its normal way may be enough for liability. Recently, there has been a 'tidal wave' of litigation for products of technology. In some cases, this stifles innovation in development of useful products. For example, a drug that works great 99.8% of the time and leads to serious harm 0.2% of the time can cost developers billions of dollars in lawsuits, much more than profits from the sale of it. Nano-bio products face this climate of liability and litigation. Statutory changes may be needed if we are to reap benefits of nanotechnology, such as tightening of regulations so mishaps do not occur; insulating manufacturers if they follow tight regulations; and creating selected compensation funds (Ziegler, 2005*).

In summary, there are several important questions regarding oversight at the nano-bio interface, including whom the public trusts, whether risk information is available and appropriately communicated, whether systems are transparent, who will be responsible for unforeseen harm, whether long-term risk issues have been analyzed, whether federal agencies have the authority and resources to appropriately regulate nanotechnology, and what the role of non-regulatory schemes will be.

Striking a balance

Society is both a supporter and watchdog of new technologies, and it strikes a balance between allowing technology to flourish and limiting it to acceptable use. Organizations play multiple roles, sometimes both promoting technology while ensuring its safety. The societal context for the nano-bio interface includes social, economic, institutional, political, and ethical issues and does not limit governance to technical risk or regulatory authority. In the past, there have been tendencies in scientific communities to argue for oversight based solely on "sound science." Yet, science alone cannot determine what level of risk is acceptable. Decisions are seldom based on science alone, especially when significant scientific uncertainty exists. Risk managers consider other factors, including social and economic ones, before making decisions.

Societal contexts come into play in setting the scope of technical risk assessments and interpreting their relevance for communities (NRC, 1996). In this model, the public is not just the recipient of risk communication, but actively involved in all facets of risk analysis and decision making. In the third session of the workshop, questions of oversight were expanded from technical risks and benefits of nanoparticles and regulatory frameworks and authorities, to societal contexts that interpret and affect nano-bio applications.

There is a need to understand public viewpoints on nanotechnology and learn more about its context from the perspective of non-experts. In recent studies on public perception of nanotechnology and trust in government, 95% of respondents did not trust government or industry to effectively manage the risks associated with nanotechnology (Cobb & Macoubrie, 2004; Macoubrie, 2005a*). In experimental issue groups where people were given information on nanotechnology, medical and general industrial applications for nanotechnology generated lower trust and that higher education levels of participants predicted lower trust. People's basis of concern was generally experience, or a "history of failed precautions." However, at the same time, people are excited about the benefits and the knowledge to be gained through nanotechnology (Cobb & Macoubrie, 2004; Macoubrie, 2005a*).

In 2005, another study was conducted to address why there is such low trust in government and industry, what people want that would increase trust, and where people are presently getting information about nanotechnology. The work also addressed whether trust was relative to specific regulatory agencies or parts of government, why there was such low trust in medical and industrial applications, and whether new information on nano-bio convergence would have influence on trust (Macoubrie, 2005b). In this study, it was found that people generally do not have knowledge of nanotechnology, but once given information, they generally have neutral or positive attitudes about it. There is increased trust in some agencies after learning about nanotechnology, but less in others. The study groups recommended the following for restoring public trust: better and more testing to discover risks; engaging the public and providing information; and adopting mandatory standards. Group participants generally thought that voluntary standards are not enough. However, 76% said that a ban on nanotechnology would be over reacting. These studies demonstrate that the public has important perspectives, ones that experts, industry, and regulators should know and consider in order for nanotechnology to succeed.

Small companies are aware of the importance of public perception and the value of proceeding with caution (Hoerr, 2005*). Nanotechnology presents unique risk characteristics, which are often unknown when work with the materials begins in a start-up setting. There are few sources to which to turn for help with risk information, yet to reduce risk to workers, industry needs to control exposure (Hoerr, 2005*). In order to do so, industry uses adherence to good manufacturing principles and safe material handling procedures. Although safety is important to industry, particularly for the sake of small companies, future regulation needs to be data-driven and cost manageable, as heavy regulatory burdens could have profound effects on small companies.

Nanotechnology has applications in other sectors, outside of pharmaceutical and biotechnology companies, such as in agriculture and food (agrifood nanotechnology). In this sector, one important social issue is power relations in the food system, which extends from suppliers to farmers; processing, distribution and marketing in industries; and finally to consumers (Thompson, 2005*). Retailers have become increasingly interested in upstream processes for inventory control and product standards (e.g. grocery chains refusing to sell genetically modified foods). Consumers are also starting to show interest in upstream processing (e.g. organic farming, animal treatment). Currently, regulation is based on the end-product and health and safety standards, but shifts may be necessary as consumers become more interested in processes in the supply chain. There will be a need to balance market power with protecting consumer rights to know and choose (Thompson, 2005*).

Another issue that was discussed at the workshop is the effects of nanotechnology on the social systems surrounding agriculture. There is evidence that technology has changed the structure of agriculture (NRC, 2002). Agricultural biotechnology was debated in this context, and as a result, there is dissatisfaction with industrial agricultural technology (Thompson, 2005*). Non-governmental organizations (NGOs) and community groups have pushed to influence directions for agricultural research that more directly serve the public (UCS, 2001). However, these concerns have not been integrated into oversight, and therefore, such groups have shifted focus to where they can challenge what is occurring, for example, through litigation and regulatory challenges. Concerns about the structure of industry, consumer choice, and ultimately, what is in the best interest of the public, will likely be prominent in public discussions about agrifood nanotechnology.

Several NGOs are optimistic about nanotechnology's promise for the environment – for cleaner or renewable energy, more efficient lighting, water filtration, and light weighting of materials – and they are eager to work with industry on realizing this promise. Yet at the same time, they are concerned about the toxicity of engineered nanomaterials (Florini, 2005*). For example, in 1998, Environmental Defense (ED) worked in collaboration with the American Chemistry Council and EPA to create a program under which chemical producers agreed to generate screening-level toxicity data for high production volume (HPV) chemicals, illustrating that voluntary initiatives can play a useful role under certain circumstances in addressing environmental concerns. The HPV Challenge was prompted in part by an ED study showing that 71% of a pilot group of HPV chemicals lacked basic screening for toxicity from what could be determined from the public record. Many argue that even more extensive information should be available for nanomaterials, in light of the novel properties that nanomaterials may exhibit (Florini, 2005*).

Steps to work on safety and regulatory challenges include more safety research funding by government and industry, effective regulations, introduction of voluntary interim standards, and meaningful stakeholder engagement. There is skepticism in the environmental law community about whether statutes, with the exception of FFDCA and TSCA, will be used for nanomaterials in the near to medium term, and there are a number of critical issues as to the use of TSCA (Florini, 2005*). It is not clear whether nanoparticles are "new" chemicals under TSCA, and thus the trigger for the Pre-Market Notification (PMN) program; what data are needed for PMN reviews; and whether current PMN exemptions are appropriate. ED proposes that an engineered nanomaterial should be considered new regardless of whether its molecular structure of formula is new, unless its chemical and physical properties are demonstrably the same as its conventional analog. Furthermore, guidance on data that should be included with a PMN is needed. Basic information on chemical characteristics, environmental fate and transport, and toxicity seem important (Florini, 2005*).

Many speakers at the workshop stressed the importance of stakeholder involvement in the design of oversight systems, including people from civil society, labor, industry, and academic researchers. Increased public understanding of nanotechnology will help facilitate public dialogs. Nanotechnology should not be put in a "black box" and thought of as having "impacts" on society. Instead, the box should open and interactions between social and ethical practices and creating knowledge at the nanoscale should be studied (Baird, 2005*). Communication about nanotechnology will impact how the public views nanotechnology and ultimately affect governance of it. Several groups are working on public communication of nanotechnology.[5]

The future

The final session of the workshop addressed the need to look toward and consider far-future applications of nanotechnology in the design of oversight systems. For example, some are now concerned that the 1986 Coordinated Framework for the Regulation of Biotechnology was not designed for the present diversity of biotechnology products and that existing laws are being twisted in strange ways (e.g. genetically engineered animals are proposed to be regulated as new animal drugs under FDA). The developers of the coordinated framework did not necessarily envision the types of products now in development.

With future applications, additional governance issues arise. Time frames for the future of nanotechnology were discussed at the workshop: mid- and longer term (Peterson, 2005*). In the mid-term, 5 years and beyond, more active nanostructures will be developed, such as sensors, actuators, and targeted drugs. It is difficult to get information in this time frame, as it is beyond most business time frames. However, the military is looking at nanotechnology in this timescale. In

---

[5] For example, see University of South Carolina Nano Science and Technology Studies: Societal and Ethical Implications. http://www.nsts.nano.sc.edu/team.html

the longer term, molecular nanosystems, not just materials or single devices, will likely emerge, and the boundaries will be limited only by what is physically or chemically possible. Longer term goals for the use of nanotechnology might also include more complete control of the structure of matter, making materials atomically precise, and designing molecular machines to do work. Health and environmental safety issues will still be important, however, concerns about privacy and surveillance will increase, as well as the use of nanotechnology for terrorism.

Applications to human enhancement will present society with fundamental social and ethical issues. There is current discussion about applications at the nanotechnology-biotechnology-information technology-cognitive (NBIC) science interface (Roco & Bainbridge, 2003). Nanotechnology could someday be used to improve senses, memory, strength, and beauty; to delay aging; or to control emotion and personality. Differences in international governance could create inequalities, as only people who can afford to travel and pay for such enhancements will benefit. Furthermore, there will be cultural differences in acceptance of these applications, and values of various societies need to be respected (Peterson, 2005*). To address future issues, guidelines for developers of nanotechnology, which are designed to address the potential positive and negative consequences in an open and accurate matter, are being formulated. The objective of the guidelines is to provide a basis for informed policy decisions by citizens and governments (Foresight, 2004).

Governance of far-future applications at the nano-bio interface also involves strategic investment in research and development. There are arguments to invest heavily in nanotechnology to solve important societal problems. The U.S. federal government is currently investing about $1 billion a year in nanotechnology, but it makes much larger investments in science that is not as ground breaking (Smith, 2005*). Federal agencies often require the techniques and concepts to be proven before large investments are made. There is the argument that big investments in nano-applications will pay off in the long run, and at the workshop, it was recommended to invest more heavily in long-term, problem-focused nanoscience (Smith, 2005*).

Conclusions and recommendations

The following conclusions and recommendations reflect themes that emerged from the workshop. However, given the diversity of participants both in the small dialog group[6] and in the audience for the public workshop[7] not all of them necessarily agree with each particular conclusion.

Defining nanotechnology

Participants at the workshop disagreed to some extent on the importance of defining nanotechnology and its applications to biology. Some argued that definitions matter because we need to attribute benefits and pitfalls to nanotechnology, and they may be necessary with respect to governance under legal statutes and regulations. However, the nanotechnology-biology interface is broad, and many believe that we should be thinking about governance more broadly. Because there are so many diverse applications, it is important to distinguish between agricultural, industrial, and medical applications, as there are different regulatory and marketing implications, as well as different public perception and acceptance landscapes.

Participants questioned whether there will there be a different set of requirements if something is defined as "nano." FDA currently regulates according to product, not process, and in this case, the definition of nanotechnology might not be important for regulation. But many argue that there should be a different set of requirements given the public claim that nanotechnology and nanoproducts are new, have distinct properties, and allow us to do special things. Many worry that developers tell the public that nanotechnology is unique and will provide great benefits, and then turn around and tell them that a special regulatory look is not necessary. This seeming contradiction

[6] The dialog group includes presenters and other participants who met on the day following the public workshop. A complete list of contributors can be found in the full report from the workshop at http://www.hhh.umn.edu/centers/stpp/nanotechnology.html

[7] 160 people from academe, trade associations, the community, industry, non-profit organizations, state and federal government, and other organizations attended the public session of the workshop.

178

may cause a loss in public confidence in nano-technology.

It is difficult to agree on a single definition for regulatory purposes, especially because every agency has to define nanotechnology in its own way based on the statutory authority it has. As a historical example, the Coordinated Framework for Biotechnology has operated for 20 years with different definitions of biotechnology in the federal agencies. However, many argued that definitions matter for other reasons – for example, so different disciplines have a common language for working with each other. Appreciating the way that words acquire meaning is important for communicating among disciplines and with the public

The group generally agreed that we should change the use of "nanotechnology" for this field to "nanotechnologies", given the vast diversity of techniques and applications. There was also general agreement that definitions are important for particular conversations, such as those at the workshop, so that participants know the focus of dialog. However, energy and resources could be misspent coming up with a single, universal definition.

Governance mechanisms

Questions about governance were the focus of the workshop, and the group considered oversight paths for nanotechnologies applied to biological systems. Every new technology goes through a phase where society has choices in oversight mechanisms, and for nanoproducts, paths for governance have not been completely figured out. There is historical precedence for defaulting to systems that are already in place, and this default is a choice for nanotechnologies. Other choices include rethinking current systems and redesigning them to better fit the nanoproducts. Regardless, many participants stressed the need for greater interactions among policymakers, toxicologists, regulators, and developers of nanomaterials to address challenges with their oversight.

For nanoproducts, jurisdictional issues are important, as there are so many products emerging in the marketplace today. Overlaps or gaps among agency authorities need to be considered and addressed, such as the lack of required pre-market testing for cosmetics and foods.

Because of delays in new rule making at the federal level, many participants stressed the need

for interim, voluntary guidelines, as it is easier to formulate them than to issue new regulations. The group discussed whether there should there be a ban on certain types of research, ones for which there are significant unknowns about the consequences. For example, in the mid 1970s with recombinant DNA technology, scientists got together at the Asilomar conference and placed some restrictions on their work until more information could be obtained. At the workshop, the question arose about whether the same caution is needed for laboratory work with nanoparticles given the associated occupational health hazards. The participants also discussed whether it is appropriate to put restrictions on research that may lead to applications that have great social consequences. Either way, interim production and use standards have a role to play, especially at the international level. Many experts at the workshop noted, however, that the public has more confidence in mandatory systems, and in the end, solid government regulation builds trust and can be industry's friend.

Specific ideas for governance from the presentations and dialog are listed below.

- Safety and toxicity issues should be discussed in open and multi-disciplinary settings before the widespread use of nanoparticles.
- New strategies in toxicology are needed to address the fact that nanomaterials have unique properties. Risk assessment paradigms may be the same, but the special properties of nanomaterials suggest that data and information needs for ensuring safety will be different.
- Governance frameworks for products of other technologies should be analyzed in order to learn from their lessons and assess their relevance to nanoproducts.
- Amending or developing new regulations and statutes is unlikely in the short, and possibly long-term; therefore, creative ways to ensure that nanotechnology is used responsibly are needed. Voluntary programs and industry standards and guidelines can provide a bridge for ensuring health and environmental safety, but they should not be considered a permanent fix, as they will not likely foster public confidence.
- EPA needs additional resources to bolster its abilities to provide oversight for nanoproducts. The agency has multiple roles of funding

research, risk assessment, and product safety review. Additional institutional capacity is needed to keep abreast of research and development in nanotechnology.

- FDA is challenged by limited statutory authority for some nanoproducts (e.g. cosmetics), and a lack of basic scientific information about nanomaterials and appropriate scientific tools to evaluate them.
- Nanomaterials might not be considered new under TSCA, if the molecular structure or formula is the same as chemicals already on the list. This could lead to gaps in oversight of health and environmental safety issues for nanomaterials. Engineered nanomaterials should be considered new regardless of whether its molecular structure of formula is "new," with the exception if their chemical and physical properties are demonstrated to be the same.
- EPA should provide nanomaterial producers with guidance on data that should be provided with PMNs, such as basic information on chemical characteristics, environmental fate and transport, and toxicity.

Social context and public engagement

Several participants stressed that public engagement in governance is needed and that there should be feedback mechanisms for public engagement early and often. The public should have input on the types of research and commercial products that are acceptable. Yet there are significant challenges to public engagement, including how to do it right and how to factor it into regulatory decisions (e.g. most legal statutes do not allow for this). Others noted that early public engagement might not always be beneficial to industry, or even society as a whole. Some argue that this has been the case for embryonic stem cells in the United States, where the technology has been highly politicized and restricted. So, if the public is engaged in decision making, society might not get answers that are best for reaping the benefits of the technology.

Public engagement in risk analysis was considered at the workshop. Technical risk assessments give information on the magnitude and types of risks, but they cannot determine what "acceptable" levels of risk are. Different publics and cultures will view the same quantitative or qualitative risk differently. Risk perception leads to various interpretations, affected by whether people understand the risk, can control it, and choose to be exposed (Slovic, 1987). Also, if there are greater rewards, people will likely accept greater risks (e.g. in the use of nanoparticles that target cancer cells).

Traditional risk assessment paradigms have been focused on the need to be science driven, or based on "sound science." There has generally been a linear progression in that the results of risk assessments are handed to risk managers who make decisions. Then, the social context comes into play during the risk communication phase. However, there is a current wave of thinking that stakeholders and citizens should be more involved in setting the questions asked by analysts and helping to interpret them in their social context (NRC, 1996). Currently, there are not good institutions for public engagement. Independent bodies of experts whom people trust and with whom they can openly communicate are needed.

Goals of public engagement in risk analysis and governance can include either better regulation in a utilitarian sense or simply choosing the right thing to do from an ethical perspective. Although public participation in setting questions and providing knowledge for risk analysis is important, the public should not be forced to argue in the context of scientific or technical risks. There are multiple issues that concern people – the notion of what is natural and what is not, corporate control of technology and society, commercialization of science, government secrecy and impacts on trust, and how technology will affect their communities. There should be opportunities for discussion of these issues, so that they do not get confused with the technical risks and benefits to human health and the environment. Finally, there are good and bad ways of doing public participation. If the public is engaged, they should either have some ability to influence the outcome or at least know that many perspectives were considered in making decisions.

Transparency and honesty in governance is needed. Mistakes have been made with previous technologies. For example, it is difficult to get information on the regulation of biotechnology products, in part due to confidential business information (CBI). This problem might be magnified with nanoproducts, where the very chemical nature could be CBI. Currently, it is difficult to find out what is coming down the pipeline in the nanotechnology industries due to intellectual property rights and CBI. There is a need to consider how to balance the benefits of intellectual property protection with better transparency in governance.

The future will be increasingly global. There are great global challenges associated with food, water, equity, the environment, and security. Nanotechnologies can play important roles in addressing these challenges. However, in the absence of institutions that can focus on societal contexts, it will be difficult to come to international agreements on the appropriate use of the technologies. Cultural differences will come into play with respect to applications for extending life, human enhancement, or privacy. Social and ethical issues not directly related to nanotechnologies will likely become associated with them. The emerging NBIC interface to enhance human performance might fly in the face of what some people believe is an acceptable limit to science and challenge the notion of what it means to be human. Application of nanotechnologies such as these might be feared or perceived negatively by the public, and all applications might become tainted. Ethical debates should distinguish among various applications, and not only include utilitarian ethics, but also intrinsic questions about playing "god," rights to know and choose, and equitable deployment of technology.

Below are several findings, conclusions, and recommendations on the social context for nanotechnologies that arose from the workshop discussions:

- Conversations about nanotechnology should not be confined to science and safety. There are other important issues in nanotechnology governance, such as the structure of industry, equity of technology deployment, life-cycle of products, consumer rights, appropriate limits of technology, and power and control over the technology.
- Although public engagement is important, the oversight process should not be driven by fear, as unnecessary heavy regulatory burdens could have profound effects on industry.
- To increase transparency, basic information on

nature and toxicity of nanomaterials should be in the public record before entering the market.
- Novel social and ethical issues will likely arise from the far-future applications of nanotechnologies, especially at the nano-bio interface. As a society, we need to begin to consider such issues, while stretching our minds about what possible applications will emerge.
- Better social institutions at local, national, and international levels are needed for diplomacy, people to relate to each other, and discussing social, ethical and political issues surrounding the nano-bio interface.

Communication and education

In studies presented at the workshop, the public seems to want more information about nanotechnology. Participants discussed needs in communication and education, such as a "citizen school" about nanotechnology. Outreach could be done through local libraries, YMCAs, science museums, and community groups. There were suggestions to get students involved in education and thinking about social issues. Citizens generally want to learn more and have the capacity to do so, but experts and policy makers need to be willing to engage with them, listen, and educate.

Right now, there is a lot of hype about the promise of nanotechnology, and developers need to be careful about what they promise that the technology can do. Hype can ultimately lead to adverse public reaction, and the public can generally see through it, especially if they are not seeing the benefits directly. There is a need to be cautious of overselling nanotechnology, and independent voices for education and communication are needed. Communication about nanotechnology will impact how the public views nanotechnology, and ultimately it will affect the governance of it.

Research agenda

Research is an integral part of steering technologies. In the workshop discussions, many pointed out that more funding is needed for research on the implications, or impacts of nanomaterials. There is also a need for applications research to be tied to implications, and the two should be integrated in the federal research agenda. One specific implications research need is long-term toxicity tests on and testing protocols for nanoproducts. Some participants suggested that a coalition of industry, NGO, government and academe should consult and jointly fund safety work.

Other research needs that were suggested include more projects on big, problem-focused nanotechnology and on societal implications:

- Implications research should be distinguished, as much as possible, from applications research and given considerably more attention and funding.
- To solve great societal problems, structural changes in research funding are needed. Funding organizations should shift some of their focus and invest larger amounts in problem-focused nanoscience, such as clean energy, water sanitation, and disaster prevention.
- Funding for research on societal contexts for nanotechnology and public engagement methods should be increased.

Concluding remarks

The workshop was successful in highlighting the various oversight opportunities and challenges at the nano-bio interface. In the author's view, it should be viewed as a step toward continuing dialog and activity on this topic. Most people want to see nanotechnologies succeed and lead to positive developments in society; however, it is important to move forward responsibly, with safety and equity in mind.

Acknowledgements

The author gratefully acknowledges all the workshop participants for their contributions and to the workshop report that forms the basis of this paper. In particular, Norris Alderson, FDA; Lynn Bergeson, Bergeson & Campbell; Jacob Finkelstein, University of Rochester; Karen Florini, Environmental Defense; Sheldon Friedlander, UCLA; Ken Hallberg, MN Nanotechnology Initiative; Robbin Johnson, Cargill, Inc.; Jane Macoubrie, Woodrow Wilson International Center for Scholars; Evan Michelson, Woodrow Wilson International Center for Scholars; Jordan Paradise, University of Minnesota; Christine Peterson, Foresight Institute; Elizabeth Wilson, University of Minnesota; and Alan Ziegler, Converging Technologies Bar Association carefully reviewed the workshop report and provided comments. The author would also like to acknowledge the support of David Y. Pui and Kenneth H. Keller, University of Minnesota, for their support and the help of Marsha Riebe, Kana Talukder and Peter VerHage of the Center for Science, Technology, and Public Policy at the University of Minnesota with organizing the workshop.

References

Alderson N., 2005. Overview of FDA's Activities at the Nano-Bio Interface, presentation at the Nanotechnology-Biology Interface: Exploring Models for Oversight, University of Minnesota, September 15.

Baird D., 2005. Philosophical Issues in Future Applications, presentation at the Nanotechnology-Biology Interface: Exploring Models for Oversight, University of Minnesota, September 15.

Bergeson L., 2005. Overview of EPA's Activities at the Nano-Bio Interface, presentation at the Nanotechnology-Biology Interface: Exploring Models for Oversight, University of Minnesota, September 15.

Cobb M. & J. Macoubrie, 2004. Public perceptions about nanotechnology: Risks, benefits and trust. J. Nanopart. Res. 6, 395–405.

Environmental Protection Agency (EPA), 2005a. Overview of Issues for Consideration by NPPTAC. Available at http://www.epa.gov/oppt/npptac/nanowgoverviewdocument20051109.pdf.

Environmental Protection Agency (EPA), 2005b. External Review Draft Nanotechnology White Paper Prepared for the U.S. Environmental Protection Agency by members of the Nanotechnology Workgroup, a group of EPA's Science Policy Council http://www.epa.gov/osa/nanotech.htm.

Finkelstein J., 2005. Establishing a relationship~between Physico-chemical and Toxicological Properties of Engineered Nanomaterials. An Approach to Assessing Risk, presentation at the Nanotechnology-Biology Interface: Exploring Models for Oversight, University of Minnesota, September 15.

Florini K., 2005. Nano-Balancing: Voluntary and Regulatory Approaches to Ensuring Safety, presentation at the Nanotechnology-Biology Interface: Exploring Models for Oversight, University of Minnesota, September 15.

Foresight Nanotech Institute, 2004. Foresight Guidelines Version 4.0: Self Assessment Scorecards for Safer Development of Nanotechnology by N. Jacobstein and G.H. Reynolds

182

Version 4.0: http://www.foresight.org/guidelines/current.html.

Friedlander S., 2005. Modern developments in nanoparticle aerosol science and technology, presentation at the Nanotechnology-Biology Interface: Exploring Models for Oversight, University of Minnesota, September 15.

Hoerr R., 2005. Nanocopoeia: A Novel Nanotechnology to Improve Drug Formulation and Delivery, presentation at the Nanotechnology-Biology Interface: Exploring Models for Oversight, University of Minnesota, September 15.

Long T.C., et al., 2006. Titanium Dioxide (P25) produces reactive oxygen species in immortalized brain microglia (BV2): Implications for nanoparticle neurotoxicity, Environ. Sci. Technol. ASAP Web Release Date: 07-Jun-2006; (Article) DOI: 10.1021/es060589n.

Macoubrie J., 2005a. Informed Public Perceptions of Nanotechnology and Trust in Government, presentation at the Nanotechnology-Biology Interface: Exploring Models for Oversight, University of Minnesota, September 15.

Macoubrie J., 2005b. Pew Charitable Trusts Projects on Emerging Nanotechnologies, Woodrow Wilson International Center for Scholars, Informed Public Perceptions of Nanotechnology and Trust in Government, September.

Michelson E., 2005. Falling Through the Cracks: Issues with Nanotechnology Oversight, presentation at the Nanotechnology-Biology Interface: Exploring Models for Oversight, University of Minnesota, September 15.

National Nanotechnology Initiative (NNI), 2005. What is Nanotechnology? http://www.nano.gov/html/facts/whatIsNano.html. Accessed November 30, 2005.

National Research Council (NRC), 1996. Understanding Risk.

National Research Council (NRC), 2002. Publicly Funded Agricultural Research and the Changing Structure of U.S. Agriculture.

Nanoscale Science, Engineering, and Technology Subcommittee (NSET), 2005. The National Nanotechnology Initiative: Research and Development Leading to a Revolution in Technology and Industry. Office of the President of the United States, March.

Oberdörster G., et al., 2005a. Nanotoxicology: An emerging discipline evolving from studies of ultrafine particles. Environ. Health Perspect. 113, 823–839.

Oberdörster G., et al., 2005b. Principles for characterizing the potential human health effects from exposure to nanomaterials: Elements of a screening strategy. Particle and Fibre Toxicology 2, 8.

Office of Science and Technology Policy (OSTP), 1986. Coordinated Framework for the Regulation of Biotechnology, 51 Fed Reg 23302.

Peterson C., 2005. Long-term Potential, Unprecedented Challenges for Governance, presentation at the Nanotechnology-Biology Interface: Exploring Models for Oversight, University of Minnesota, September 15.

Roco M.C., 2004. Nanoscale Science and Engineering: Unifying and Transforming Tools, 50(5) AIChE 890–897, 895–96.

Roco M.C., 2005. The emergence and policy implications of converging new technologies integrated from the nanoscale. J. Nanopart. Res. 7, 129–143.

Roco, M.C. & W.S. Bainbridge, (eds.), 2003. Converging Technologies for Improving Human Performance. Boston: Kluwer Academic Publishers.

Savage N., 2005. Nanotechnology & EPA, presentation at the Nanotechnology-Biology Interface: Exploring Models for Oversight, University of Minnesota, September 15.

Slovic P., 1987. Perception of risk. Science 236, 280–285.

Smith R., 2005. Normal Science or Problem Solving? Getting Our Priorities Straight, presentation at the Nanotechnology-Biology Interface: Exploring Models for Oversight, University of Minnesota, September 15.

Taton A., 2005. The Nano-Bio Interface: An Overview, presentation at the Nanotechnology-Biology Interface: Exploring Models for Oversight, University of Minnesota, September 15.

Thompson P., 2005. Social and Ethical Implications in Agrifood Applications, presentation at the Nanotechnology-Biology Interface: Exploring Models for Oversight, University of Minnesota, September 15.

Union of Concerned Scientists (UCS), 2001. The Costs and Benefits of Industrial Agriculture. Backgrounder. http://www.ucsusa.org/food_and_environment/sustainable_food/costs-and-benefits-of-industrial-agriculture.html.

Untereker D., 2005. Presentation at the Nanotechnology-Biology Interface: Exploring Models for Oversight, University of Minnesota, September 15.

Walker L., 2005. Nanotechnology for Agriculture, Food, and the Environment, presentation at the Nanotechnology-Biology Interface: Exploring Models for Oversight, University of Minnesota, September 15.

Ziegler A., 2005. The Nano-Bio Interface: Enterprise Liability and Insurance Risk, presentation at the Nanotechnology-Biology Interface: Exploring Models for Oversight, University of Minnesota, September 15.

Journal of Nanoparticle Research (2007)

Future conferences and symposia (January 2007–)

January 16–19, 2007
2nd Annual IEEE International Conference on Nano/Micro Engineered and Molecular Systems,
Imperial Queen's Park Hotel, Bangkok, Thailand. Contact: http://www.ieee-nems.org

February 25–March 1, 2007
Nanomaterials Symposia, TMS 2007 Annual Meeting and Exhibition, Orlando, Florida. Contact:
http://www.tms.org

April 9–13, 2007
Sessions J, Z, AA, DD, EE, FF, GG, HH, II, and JJ, Materials Research Society Spring Meeting,
San Francisco, California. Contact: http://www.mrs.org/spring2007

April 16–19, 2007
3rd International Nanotechnology Conference on Communications and Cooperation (INC3), Brussels,
Belgium. Contact: http://www.imec.be/inc3/Welcome.html

June 7–9, 2007
5th International Workshop on Electrodeposited Nanostructures (EDNANO 2007), National Institute of
Research & Development for Technical Physics, Iasi, Romania. Contact:
http://www.szfki.hu/ednano/EDNANO5main.html

June 10–15, 2007
Nanocrystallization Session, Third International Conference on Recrystallization and Grain Growth (ReX
& GG III), Shilla Jeju Hotel, Jeju Island, Korea. Contact: http://www.rex-gg-2007.org

August 29–September 1, 2007
3rd International Symposium on Nanotechnology, Occupational and Environmental Health,
International Convention Center, National Taiwan University Hospital, Taipei, Taiwan.
Contact: http://www.cce.umn.edu/conferences/nanotechnology/

September 16–20, 2007
Engineering Conference International Meeting on Nanofluids: Fundamentals and Applications,
Copper Mountain, Colorado. Contact: http://www.engconfintl.org/7ax.html